SMP **interact**

for GCSE

Teacher's guide to book I2 part A

PATHFINDER EDITION

Contents

The following people contributed to the writing of the SMP Interact Key Stage 4 materials.

Benjamin Alldred
Juliette Baldwin
Simon Baxter
Gill Beeney
Roger Beeney
Roger Bentote
Sue Briggs
David Cassell
Ian Edney
Stephen Feller
Rosemary Flower
John Gardiner
Colin Goldsmith
Bob Hartman
Spencer Instone
Pamela Leon
John Ling
Carole Martin
Lorna Mulhern
Mary Pardoe
Paul Scruton
Susan Shilton
Caroline Starkey
Liz Stewart
Biff Vernon
Jo Waddingham
Nigel Webb
Heather West

Others, too numerous to mention individually, gave valuable advice, particularly by commenting on and trialling draft materials.

Editorial team: David Cassell, Spencer Instone, John Ling, Paul Scruton, Susan Shilton, Heather West

Project administrator: Ann White

Design: Pam Alford, Melanie Bull, Nicky Lake, Tiffany Passmore, Martin Smith

Project support: Carol Cole, Pam Keetch, Jane Seaton, Cathy Syred

PUBLISHED BY THE PRESS SYNDICATE OF THE UNIVERSITY OF CAMBRIDGE
The Pitt Building, Trumpington Street, Cambridge, United Kingdom

CAMBRIDGE UNIVERSITY PRESS
The Edinburgh Building, Cambridge CB2 2RU, UK
40 West 20th Street, New York, NY 10011-4211, USA
477 Williamstown Road, Port Melbourne, VIC 3207, Australia
Ruiz de Alarcón 13, 28014 Madrid, Spain
Dock House, The Waterfront, Cape Town 8001, South Africa

http://www.cambridge.org

First published 2002

Printed in Italy by Rotolito Lombarda
Typeface Minion *System* QuarkXPress®

A catalogue record for this book is available from the British Library

ISBN 0 521 01309 7 paperback

Typesetting and technical illustrations by The School Mathematics Project

No solutions to the questions were provided or approved by AQA, and the solutions provided may not constitute the only ones. Edexcel accepts no responsibility whatsoever for the accuracy or method of working in the answers given.

Introduction

Classroom organisation and teaching style

Activities which are intended to be led by the teacher are shown by a solid marginal strip.

Often in such cases there will be little, if any, corresponding material in the student's book.
The activity will be described in the teacher's guide.

Each section of the teacher's guide starts with an overview of the work.
In this overview, activities which are intended to be teacher-led are shaded.

Some activities, although not teacher-led, are likely to need a lot of teacher support (for example, to go over the results of a group activity to ensure that the main points have been understood). These activities are indicated by a broken marginal strip.

This symbol appears where a calculator is not intended to be used.

Assessment

Each substantial unit of work finishes with questions for self-assessment.

1 Handling secondary data

This work will help students analyse and interpret data that has been collected and organised by someone else.

Section A introduces students to skills involved in analysing secondary data.

This is followed by section B: Ideas for projects using secondary data.

It is expected that students choose one of them after working through section A.

A *Drawing conclusions from data* (p 4)

◊ In this section, students are introduced to some important skills involved in interpreting secondary data.

These skills are illustrated by examples with follow-up questions. The questions can be used for class discussion or individual work. The whole section is intended as a preparation for section B, where students have a choice of projects.

◊ Sometimes there are discrepancies in tables of secondary data. Because of rounding, individual amounts do not always add up to totals given. For example, in the table on page 4, the total of all modes of transport for 1995 is one less than the total of the given figures. Similarly, percentages may not add up to 100%. The differences are usually small and of little significance but students should be aware of it.

◊ Data is often given for the UK or Great Britain. Great Britain includes only England, Scotland and Wales whereas the UK also includes Northern Ireland. For example the transport data on page 4 is about Great Britain but the population data on page 10 is about the UK. Britain is sometimes used in this work as shorthand for Great Britain.

B *Ideas for secondary data projects* (p 11)

For many of these projects, a spreadsheet package is ideal for analysing the data and it may be worth having data on file prior to students beginning their investigations.

B1 ***Countries of the world*** (p 11)

◊ The notes for questions A19 to A23 are relevant here.

◊ A useful website is http://www.odci.gov/cia/publications/factbook/

B2 ***Travelling around*** (p 12)

◊ Where possible, data is given for years 1950, 1955, 1960, … . Sometimes data for these years is not available and data for an adjacent year is given. Population data is available every ten years as shown in Table 8. Students need to consider how estimates of data for other years could be found.

Note that the population in Table 8 is for Britain and is slightly different from the UK data given on page 10.

◊ Discuss the variety of units used in the tables and make sure students know that a vehicle or passenger kilometre is a kilometre travelled by a vehicle or passenger.

Students often have difficulty in interpreting, for example, 12.5 million as 12.5 × 1 000 000 = 12 500 000.

In line with common practice 1 billion = 1000 million.

◊ To familiarise students with the data, the tables could be used initially as the basis for some work on reading and interpreting tables: for example, Questions 1 to 10.

◊ In analysing the data, there is likely to be considerable use of percentages and compound measures such as 'cars per kilometre'. You may wish to make sure that students are confident in these areas before they begin their investigations.

◊ Students may try to answer a question using the data from one table only. For example, in trying to decide if our roads are becoming more congested, they may consider the data in Table 1 but fail to consider the data in Table 5. Encourage students to think about the possible relevance of all the data provided.

◊ Students who have difficulty in interpreting data from more than one table could choose one table, decide on the best way to display the data and draw some conclusions.

◊ Students who investigate 'Who is most at risk on our roads?' could produce a road safety poster aimed at our most vulnerable road users.

B3 ***Growing old*** (p 16)

◊ In addition to the 'expectation of life at birth' i.e. the average age people live to, the first table also tells us how long we can expect to live on average having already reached a certain age. So, for example, having reached the age of 80 in 1996 you could expect to live another 6.8 years on average.

◊ In the second table, you may need to emphasise that all the figures are percentages except those in the last column which are actual population figures (in millions).

Students often have difficulty in interpreting, for example, 55.9 million as 55.9 × 1 000 000 = 55 900 000.

◊ To familiarise students with the data, the tables could be used initially as the basis for some work on reading and interpreting tables: for example, Questions 1 to 10.

◊ In trying to decide what the population will be like in 100 years time, students could start by considering how projections for the year 2001 and beyond might have been arrived at for each table.

◊ In making comparisons and looking at trends, may use a variety of diagrams including line graphs, bar charts, composite bar charts and pie charts. Care will be needed with scales for axes and rounding.

◊ All data on the data sheet comes from Social Trends available as an annual book or on CD Rom from the Office of National Statistics.

◊ A useful website is http://www.doh.ie/stats/bindex.htm

B4 *Work, work, work* (p 18)

◊ Discuss the units and degrees of accuracy used in the tables. For example, as the units in Table 1 are thousands then the number of men employed in 1925 was 8,717,000 (not 8,717) and this has been rounded to the nearest thousand.

◊ Discuss the type of information in the tables. Sometimes numbers and sometimes percentages are given.

◊ To familiarise students with the data, the tables could be used as the basis for some work on reading and interpreting tables: for example, questions 1 to 10.

◊ In analysing the data, there is likely to be considerable use of percentages. You may wish to make sure that students are confident in this area before they begin their investigations.

◊ Students need not consider every table in their investigations. Some may wish to restrict their study to Table 4 for example and look solely at the type of work people have done over the years.

◊ Some students may notice that the total numbers of employees in tables 1 and 4 do not match. Discuss possible reasons for this (perhaps Table 1 does not include self-employed or the figures may refer to different times of the year).

◊ In Table 3, separate figures for England were not included.

B5 *Trouble on the terraces* (p 21)

◊ The notes under the data give the changes in league structures which might be important if students do their own research.

◊ The first, second and fourth columns of figures refer to English football and rugby organisations. The word English or England does not appear in their titles.

◊ From 1986/87 data in the first and second columns includes promotion and relegation play-off matches. This will not have affected the average attendances too much.

◊ To familiarise students with the data, the tables could be used as the basis for some work on reading and interpreting tables: for example, questions 1 to 5.

◊ In drawing graphs and charts, students may need to round data sensibly before plotting attendances.

◊ The notes under Sunday/Monday newspapers often give detailed information on football matches. One example is:

Current newspaper data could be used to provide additional information on attendances.

◊ Other investigations could be:

- Do teams in higher divisions score more goals?
- Are more goals scored towards the end of matches or the beginning?

◊ The data comes from Social Trends available as an annual book or on CD Rom from the Office of National Statistics. Other good sources of football statistics are the Rothman's Yearbook and other similar publications.

◊ A useful website for B6 is http://www.risc.uni-linz. ac.at/non-official/rsssf/nersssf.html

B6 ***Rain check*** (p 22)

◊ To familiarise students with the data, the tables could be used as the basis for some work on reading and interpreting tables: for example, questions 1 to 5.

◊ Emphasise that the figures are all means. For example, in London 7°C is given as the maximum temperature for January. This does not imply that 7°C was the highest temperature in January but that the mean of the highest daily temperatures was 7°C.

For question 5, some students may give the rainfalls as 2.3 and 17.8 mm respectively. The total of the figures in the rainfall column is 17.8 mm. However, the figures are means so they cannot be added to find the total rainfall for part (b).

◊ The data in the tables were compiled from more detailed data giving daily figures for 1991 provided by the Meteorological Office.

B7 ***Titanic*** (p 25)

◊ To familiarise students with the data, the tables could be used as the basis for some work on reading and interpreting tables: for example, questions 1 to 5.

◊ 'Steerage' was the part of the ship where passengers travelled at the cheapest rate.

A Drawing conclusions from data (p 4)

A1 255 billion km or 255 000 000 000 km

A2 6 000 000 000 km

A3 Bus and coach, pedal cycle, motorcycle, rail

A4 You cannot say for certain the more people were flying. It could have been the same number but each flying further (although this is unlikely).

A5 (a) 5900 km (b) about 11 500 km

A6 (a) (i) 13.2% (ii) 5.4%
(b) It went down

A7 It went up from 86.4% to 93.9%.

A8 It went down from 29.8% to 6.7%.

A9 You can compare the countries by looking at the number of doctors per million people:
Brazil 85.0 Egypt 183.0 Kenya 90.0
Looked at in this way, Egypt is better off than either Brazil or Kenya, though each country's medical needs and the quality of medical training are also factors.

A10 (a) Vandalism went up between 1981 and 1993, but then fell between 1993 and 1995.
(b) You would need to know the number of households.

A11 13.0%

A12 46.3%

A13 81.4%

A14 22.5%

A15 Men: 60% full-time, 40% part-time
Women: 36.7% full-time, 63.3% part-time

A16 She was incorrect. We do not know what proportion of drivers overall drive after drinking. It may be that the relatively small number of drinking drivers involved in accidents is in fact a high proportion of all drinking drivers.
To draw a conclusion we need to know (i) what proportion of drinking drivers are involved in accidents, and (ii) what proportion of non-drinking drivers are involved in accidents.

A17 There are several reasons why the article is inaccurate:
Some offences did not count as vandalism before 1982. The number of offences is given, but not the size of the population. If population increases, then we would expect a corresponding rise in cases of crime.
The spaces between the bars in the bar chart do not show that the first two bars are 11 years apart. So the increase could have been more gradual than appears.

A18 It allows the reader to check that your data is accurate and comes from a reliable source.

A19 To compare crowdedness you can calculate the population per square km.
For example:
Zaire 17.6 Brazil 17.3
UK 238.8 Netherlands 374.6

A20 Brazil has more main roads than the UK, but it is also a much larger country with a larger population.
You can calculate the length of main roads per square kilometre (Brazil 0.00177, UK 0.0512) or per million people (Brazil 102.5, UK 214.4). Either way, Brazil is not so well off for main roads.

*A21 (a) The statement is correct.

(b) The statement is incorrect. You cannot just add the 'doctors per million' of the two countries. To find the number of doctors per million of the combined country, you would have to work out how many doctors there are in each country, add them and then divide by the total population in millions.

Just adding gives 1082 + 856 = 1938 doctors per million.
However, Ecuador has about 1082 × 11 = 11902 doctors and Peru has about 856 × 22.5 = 19 260 doctors.
The combined population is 11 + 22.5 = 33.5 million and the total number of doctors is 11 902 + 19 260 = 31 162.
This gives 31 162 ÷ 33.5 ≈ 930 doctors per millions for the combined country.

(c) The statement is incorrect. You cannot just find the average of the two literacy rates. You would have to work out the number of literate adults in each country, add them and find this as a percentage of the total population. Since the number of adults in each country is not given, it is impossible to find the literacy rate for the combined country from these figures alone.

A22 (a) 32.8% of 44.03 million = 14.53 million

A23 You cannot find the number of males aged 65+ because you do not know what percentage of the males is in each age group.

B2 *Travelling around* (p 12)

1 1941 pedestrians

2 69 533 kilometres

3 1960

4 252 thousand or 252 000 road accidents

5 55.5 million or 55 500 000

6 21.9 million or 21 900 000 households

7 (a) 4900 thousand or 4 900 000 private cars

(b) 14 660 thousand or 14 660 000 private cars
Students' descriptions, for example:
- The number increased by 9760 thousand or 9 760 000
- The number increased by about 10 million
- There were about 3 times as many cars in 1980 as there were in 1960

8 Students' descriptions, for example:
- The number of kilometres decreased by 15.5 billion or 15 500 000 000
- The number of kilometres decreased by about 78%
- There were between 4 and 5 times as many miles travelled in 1950 as there were in 1970

9 (a) About 58% (b) About 76%

10 (a) About 6% (b) About 19%

B3 *Growing old* (p 16)

1 75.9 years in total (or 65.9 years more)

2 3.7%

3 58 006 thousand or 58 006 000

4 Scotland

5 31.6 more years

6 21.6%

7 The percentage more than doubles. The Government may be concerned at the cost of old age pensions or health care for example.

8 About 83%

9 About 2012

10 In 1993, the % of males was 2.8% higher than the % of females but in 2052 the gap has narrowed to 1.7%

B4 *Work, work, work* (p 18)

1 14 719 thousand or 14 719 000

2 1980

3 10.0%

4 1902 thousand or 1 902 000

5 74%

6 1955

7 1935

8 35% or 34·8%

9 (a)

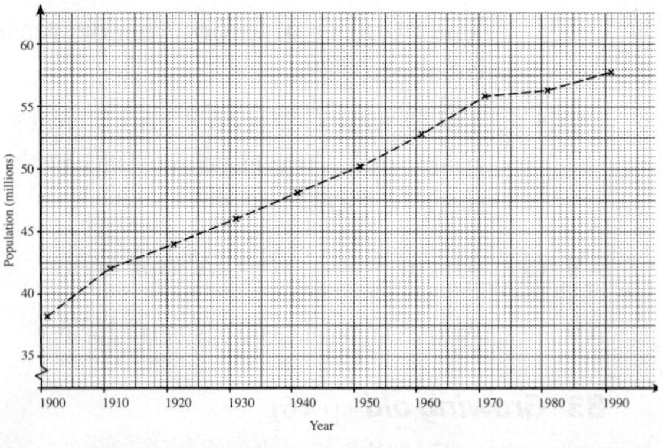

(b) (i) About 50 million or 50 000 000

(ii) About 57 million or 57 000 000

(c) (i) About 28% (ii) About 14%

(d) (i) About 21% (ii) About 17%

10 (a) 1.5% (b) Students' explanations

B5 *Trouble on the terraces* (p 21)

1 21 622

2 1971/72, 1976/77, 1989/90

3 1986/87

4 1986/87 was worst for England
1981/82 was worst for Scotland

5 About 17%

B6 *Rain check* (p 22)

1 June

2 Birmingham

3 Plymouth, London and Valley all have an average maximum temperature of 7°C. However Plymouth has the highest average minimum temperature. So, according to these figures, Plymouth was the warmest place in January.

4 August and September

5 (a) 71.3 mm (b) 539.9 mm

B7 *Titantic* (p 25)

1 (a) 2201 passengers
(b) 1490 died

2 (a) 109 children (b) About 50% died

3 About 2 of the men on board were crew

4 76%

5 66%

2 Inequalities 1

Practice booklet pages 3 to 4

A *Single inequalities* (p 26)

◊ Discuss the symbols <, >, ≤, ≥ and the equivalence of (for example) $x < 3$ and $3 > x$. Explain the conventions used in representing simple inequalities on a number line. Students can then try to match up as many of the statements and diagrams A to J as they can.

The correct matches are A with J, E with C, I with B and G, D with H. F is unmatched.

A1 Some students may feel, for example in part (g), that $0.19 \leq 0.2$ is false because $0.19 \neq 0.2$. It might help to point out that $0.19 \leq 0.2$ means that either $0.19 < 0.2$ **or** $0.19 = 0.2$ and since $0.19 < 0.2$ is true then the whole statement $0.19 \leq 0.2$ is true.

A3 Point out that here (as in later questions) a quick sketch of the number line, not an accurate drawing, is all that is required.

B *Combined inequalities* (p 27)

◊ Students might find it helpful to sketch separate diagrams, as shown in the introduction, as an aid to drawing the combined diagram.

This method will show clearly when there is no overlap and hence when separate inequalities cannot be combined in this way.

◊ Emphasise that a combined inequality means that the variable must satisfy both inequalities–not just one or the other.

◊ Students can again try to match up as many of the statements and diagrams A to J as they can.

The correct matches are A with B and G, E with D and J, I with C and F. H is unmatched.

◊ You may wish to draw students' attention to the fact that some apparently combined inequalities are contradictory, such as $5 \leq x \leq 1$.

C *In words and symbols* (p 29)

TG

Emphasise that you need to be precise in the definition of letters, e.g. $b \geq 5$ does not mean 'black biros are greater than or equal to 5'. This is too vague and could refer to the length or weight of black biros for example. In this case you need to state that b stands for the **number** of black biros in my bag.

A *Single inequalities* (p 26)

A1 (a) T (b) T (c) T (d) T
(e) T (f) T (g) F (h) F
(i) T (j) F (k) F (l) T

A2 (a) $-20, 1, \frac{2}{5}$ (b) $\sqrt{10}, \frac{13}{4}$

A3 (a)
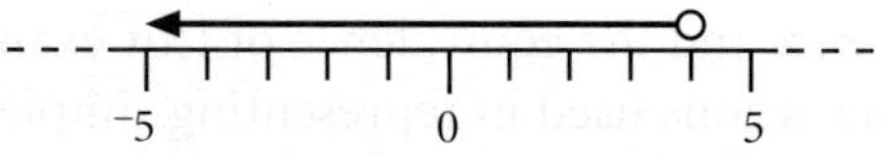
(b)
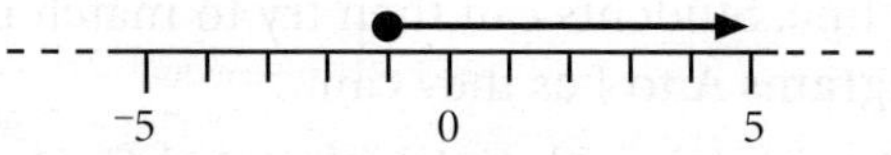
(c)
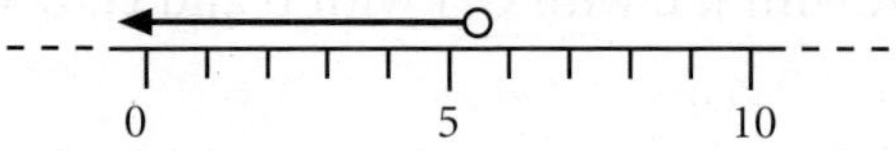
(d)
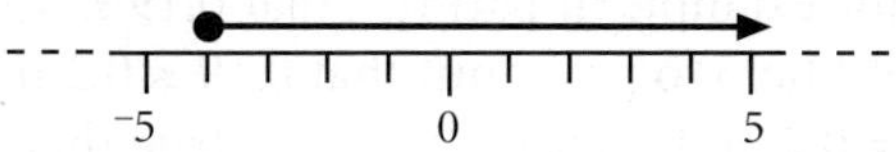

A4 (a) $x \geq -5$ (b) $x < 0$
(c) $x \leq 2.5$ (d) $x > -2$

***A5** (a) Sometimes (b) Always
(c) Always (d) Never
(e) Always (f) Sometimes
(g) Always (h) Never

B *Combined inequalities* (p 27)

B1 (a)
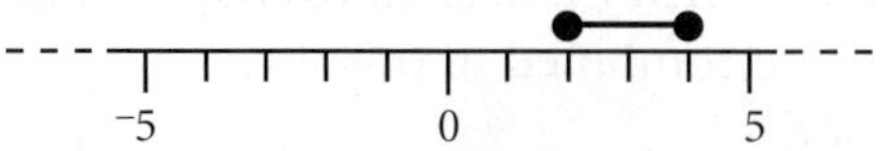
(b)
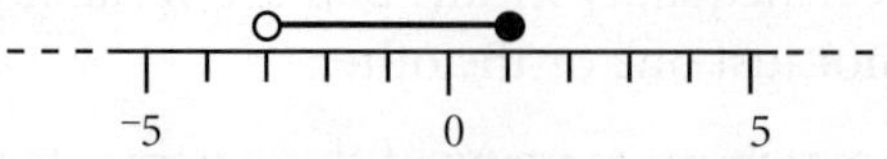
(c)
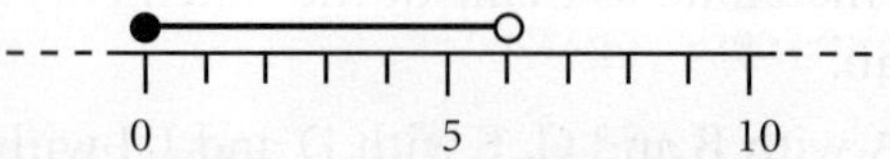
(d)
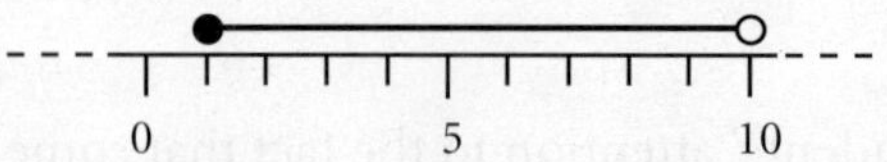

B2 (a) $\sqrt{4}, \sqrt{24}$ (b) $\sqrt{4}, -2, 0$

B3 (a) $-3 \leq x \leq 4$ (b) $-5 < x \leq 0$
(c) $0 < x < 5$ (d) $-5 \leq x < -1$

B4 0, 1, -1.5

B5 0, 1, 2, 3

B6 Any five numbers between -1 and 2 (they may include -1, but not 2)

B7 (a) -2, 0 and 1 000 000
(b) 2.1, 0, 1.5 and 0.5

B8 9, -9, 9.9 and 0

B9 $-4, -3, -2, -1, 0, 1, 2, 3$ and 4

B10 -4 and 4

B11 Any integer less than or equal to -5

B12 23 and 29

B13 1002, 1005 and 1008

B14 Any three of 2, 3, 4, 5 and 6

***B15** Any three of 5, 6, 7, 8, 9, 10 and 11

***B16** -1, 0 and 1

C *In words and symbols* (p 29)

C1 (a) $t \leq 4$ (b) $t < -18$ (c) $t \geq 24$

C2 $w \leq 7$

C3 (a) $n \geq 20$ (b) $s \leq 35\,000$

C4 The student's sentences

Test yourself (p 30)

T1 (a) F (b) T (c) F
(d) T (e) T

T2 (a)
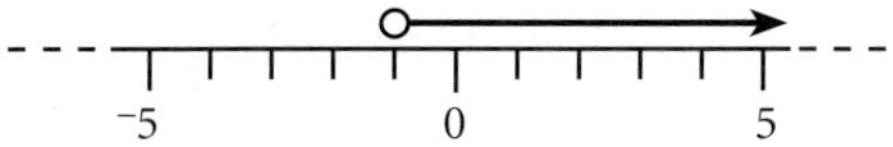

(b)
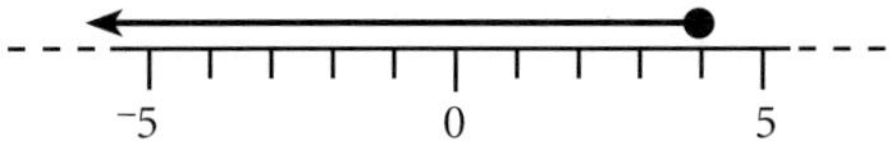

(c)
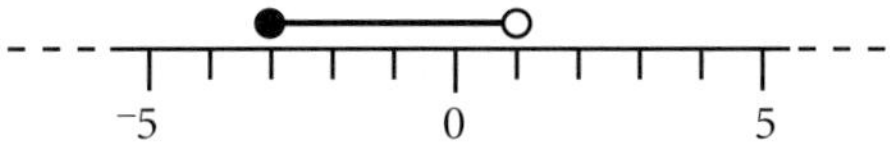

(d)
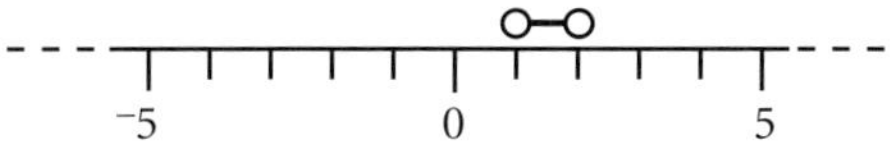

T3 (a) $x \geq 2$ (b) $x < 3$
(c) $x > 0$ (d) $x < {}^-2$

T4 (a) $^-1$ and 0
(b) $^-5$, $^-4$, $^-3$, $^-2$ and $^-1$
(c) $^-3$, 3, $^-4$ and 4

T5 (a) $0 < x < 4$ (b) ${}^-3 \leq x \leq 4$
(c) $1 \leq x < 5$ (d) ${}^-4 < x \leq 3$

T6 2, 3, 4 and 5

T7 $^-1$, 0, 1, 2 and 3

T8 1, 2, 3, 4 and 5

T9 Five from $^-3$, $^-2$, $^-1$, 0, 1, 2 and 3

T10 (a) $p \leq 25$ (b) $c > 20\,000\,000$
(c) $x \geq 50\,000$ (d) $w < 5$

Practice booklet

Section A (p 3)

1 (a) False (b) False (c) True
(d) False (e) False (f) True
(g) True (h) False

2 (a) $\sqrt{7}$, 0.7, $^-7$, 3 (b) $\frac{25}{3}$, $\sqrt{50}$

3 (a)
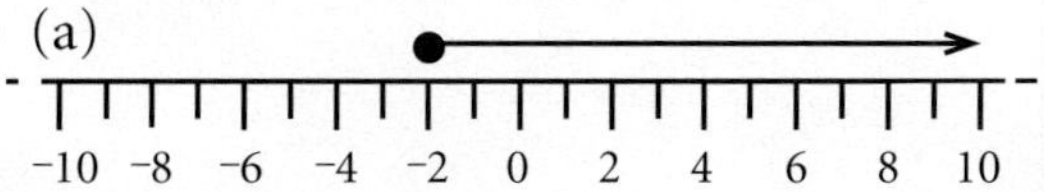

(b)
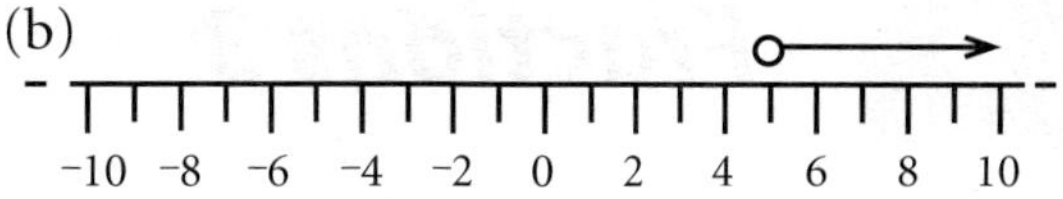

(c)
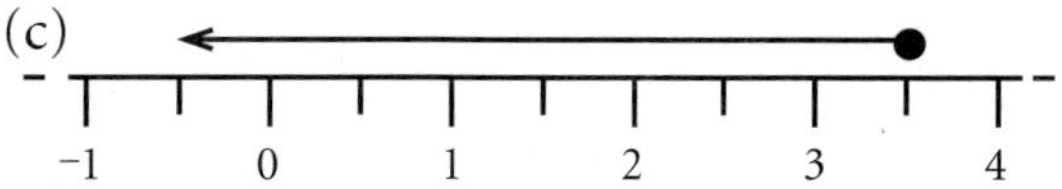

4 (a) $x > {}^-4$ (b) $x \leq 6$
(c) ${}^-4 \leq x < 6$

5 (a) Always (b) Sometimes
(c) Always (d) Never
(e) Always (f) Never

Section B (p 4)

1 (a)
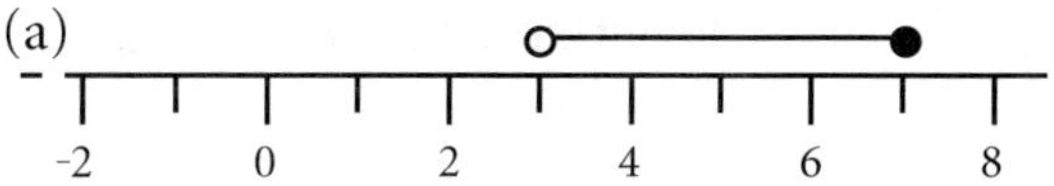

(b)
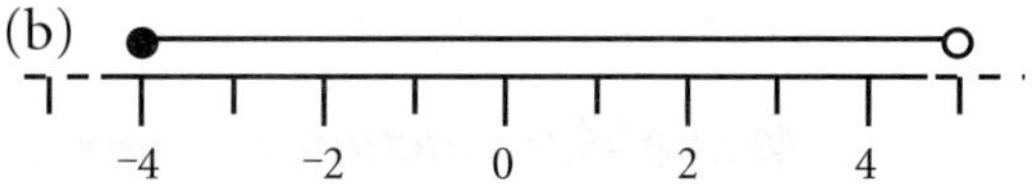

(c)
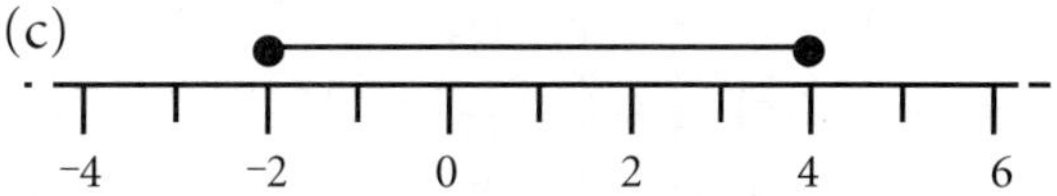

2 (a) 0, $\frac{3}{2}$, $\sqrt{8}$, $^-0.5$, $^-1$ (b) 0, $^-6$, $^-0.5$, $^-1$

3 (a) ${}^-4 \leq x < 6$ (b) ${}^-4 \leq x \leq 8$
(c) ${}^-6 < x < 4$

4 Any 5 numbers between $^-3$ and 1 (including 1 but not $^-3$)

5 $^-3$, $^-2$, $^-1$, 0, 1, 2, 3

6 Any two of $^-5$, $^-4$, 4, 5

7 Any integer less than or equal to $^-3$

8 Any four values between and including $^-1$ to 4

Section C (p 4)

1 (a) $p \leq 500$ (b) $s \geq 8$
(c) $t < 50$ (d) $s > 4000$

2 Student's sentences

3 Fractions 3

This unit completes work on the four operations with fractions.

Practice booklet page 5

A *Multiplying fractions: review* (p 31)

A1 (a) $\frac{1}{3} \times \frac{1}{4} = \frac{1}{12}$ (b) $\frac{1}{3} \times \frac{5}{6} = \frac{5}{18}$
(c) $\frac{3}{8} \times \frac{2}{3} = \frac{6}{24}$ $(= \frac{1}{4})$
(d) $\frac{3}{5} \times \frac{3}{4} = \frac{9}{20}$

A2 (a) $\frac{1}{2}$ (b) $\frac{3}{10}$ (c) $\frac{5}{8}$
(d) $\frac{4}{15}$ (e) $\frac{6}{25}$

A3 (a) $\frac{9}{8}(= 1\frac{1}{8})$ (b) $\frac{8}{9}$ (c) 2
(d) $\frac{15}{8}(= 1\frac{7}{8})$ (e) $\frac{15}{4}(= 3\frac{3}{4})$

A4 (a) 1 (b) 1 (c) 1
(d) 1 (e) 1

B *Reciprocals* (p 32)

B1 (a) $\frac{1}{3}$ (b) 3 (c) $\frac{1}{5}$ (d) $\frac{5}{2}$
(e) $\frac{6}{5}$ (f) 8 (g) $\frac{7}{5}$ (h) $\frac{4}{5}$
(i) $\frac{1}{8}$ (j) 1

B2 (a) $\frac{3}{2}$ (b) $\frac{2}{3}$

B3 (a) $\frac{2}{7}$ (b) $\frac{3}{4}$ (c) $\frac{4}{9}$
(d) $\frac{5}{7}$ (e) $\frac{3}{8}$

C *Dividing a whole number by a fraction* (p 32)

C1 9

C2 24

C3 (a) 8 (b) 15 (c) 16
(d) 20 (e) 18

C4 (a) 10 (b) 5

C5 (a) 24 (b) 8 (c) 18
(d) 9 (e) 10

C6 (a) $\frac{15}{2}$ (b) $\frac{8}{3}$ (c) $\frac{20}{3}$
(d) $\frac{4}{3}$ (e) $\frac{40}{3}$

C7 (a) 24 (b) 12

C8 8 hours

C9 25 kg

D Dividing a fraction by a fraction (p 34)

D1 (a) $\frac{10}{9}$ (b) $\frac{10}{3}$ (c) $\frac{8}{9}$
(d) $\frac{3}{4}$ (e) $\frac{6}{5}$

D2 (a) $\frac{5}{6}$ (b) $\frac{2}{9}$ (c) $\frac{8}{15}$
(d) $\frac{10}{9}$ (e) $\frac{9}{8}$

E Substituting into algebraic expressions (p 34)

E1 (a) 2 (b) 3 (c) $2\frac{1}{2}$ (d) 0
(e) $4\frac{1}{2}$ (f) $\frac{1}{6}$ (g) 12 (h) 12
(i) $\frac{3}{2}$ (j) $\frac{2}{3}$

E2 (a) 4 (b) 6 (c) 1 (d) $2\frac{1}{2}$
(e) $\frac{2}{3}$ (f) $\frac{1}{2}$ (g) $\frac{3}{2}$ (h) $\frac{4}{3}$
(i) $\frac{8}{9}$ (j) $\frac{9}{8}$

E3 (a) $\frac{13}{10}$ or $1\frac{3}{10}$ (b) $\frac{3}{10}$ (c) $\frac{2}{5}$
(d) $\frac{8}{5}$ (e) $\frac{5}{8}$

Test yourself (p 34)

T1 (a) $\frac{1}{4}$ (b) 5 (c) $\frac{3}{2}$ (d) $\frac{5}{8}$

T2 (a) 18 (b) $\frac{18}{5}$ (c) 4
(d) $\frac{4}{5}$ (e) $\frac{15}{16}$

Practice booklet

Section A (p 5)

1 (a) $\frac{1}{6}$ (b) $\frac{1}{3}$ (c) $\frac{3}{8}$
(d) $\frac{3}{5}$ (e) $\frac{1}{4}$

2 (a) $\frac{9}{4}(= 2\frac{1}{4})$ (b) 1 (c) $\frac{10}{3}$
(d) $\frac{16}{9}$ (e) 3

Section B (p 5)

1 (a) $\frac{1}{7}$ (b) 7 (c) $\frac{7}{2}$
(d) $\frac{9}{4}$ (e) $\frac{5}{8}$

2 (a) $\frac{4}{5}$ (b) $\frac{2}{5}$ (c) $\frac{2}{9}$
(d) $\frac{3}{4}$ (e) $\frac{3}{10}$

Section C (p 5)

1 (a) 14 (b) 28 (c) 30
(d) 72 (e) 25

2 (a) 15 (b) 18 (c) 25
(d) 16 (e) $\frac{20}{3}$

Section D (p 5)

1 (a) $\frac{4}{9}$ (b) $\frac{9}{10}$ (c) $\frac{6}{5}$
(d) $\frac{5}{2}$ (e) $\frac{7}{6}$

Section E (p 5)

1 (a) $\frac{3}{2}$ (b) 8 (c) $\frac{1}{10}$
(d) $\frac{5}{8}$ (e) $\frac{8}{5}$

2 (a) $\frac{15}{4}$ (b) $\frac{8}{5}$ (c) $\frac{3}{5}$
(d) $\frac{15}{16}$ (e) $\frac{16}{15}$

4 Locus

Essential	**Optional**
Compasses, squared paper, sheet P45	OHP
Practice booklet pages 6 to 7	

A *The distance from ...* (p 35)

Compasses

◊ These ideas can be introduced on the board or on an OHP with students coming out to mark points that satisfy each condition or to draw the required locus. You can extend the your introduction to include loci with more than one condition.

B *Nearer, further, the same distance ...* (p 38)

Compasses, squared paper

◊ If students at first do not see that the locus of points equidistant from points A and B is the perpendicular bisector of the line AB, then mark A and B on the board/OHP and invite them to mark further points one by one and decide whether each point is closer to A or to B. Those closer to A can be ringed in one colour while those closer to B can be ringed in a different colour. As points are added, ask the class if they can see where the boundary between the two colours will be and what they can say about points on that boundary.

Some may recall the ruler and compasses construction for the perpendicular bisector of the line AB. It will need to be demonstrated to those who do not.

◊ The locus of lines equidistant from *l* and *m* consists of two lines, each bisecting the angle between *l* and *m*.

Again if students find this difficult, invite them to mark points one by one and decide whether each point is closer to *l* or to *m*. Those closer to *l* can be ringed in one colour while those closer to *m* can be ringed in a different colour. As points are added, ask the class if they can see where the boundary between the two colours will be and what they can say about points on that boundary.

For those who do not recall the ruler and compasses construction for an angle bisector you will need to arrange a demonstration.

An interesting feature is that the two lines forming this locus are at right angles. You can ask students to prove how this must be the case, whatever the angles between *l* and *m*.

C ***Mixed questions*** (p 39)

Compasses, sheet P45

Many diagrams in these answers are not drawn to scale.

A ***The distance from ...*** (p 35)

A1

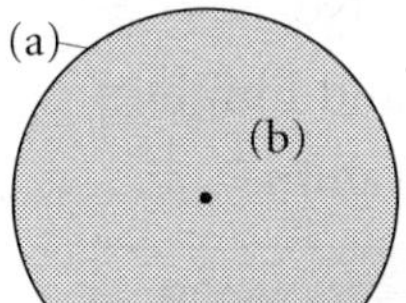

A2

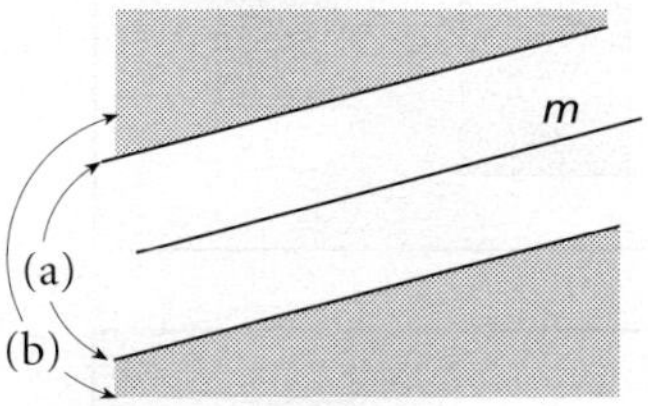

A3

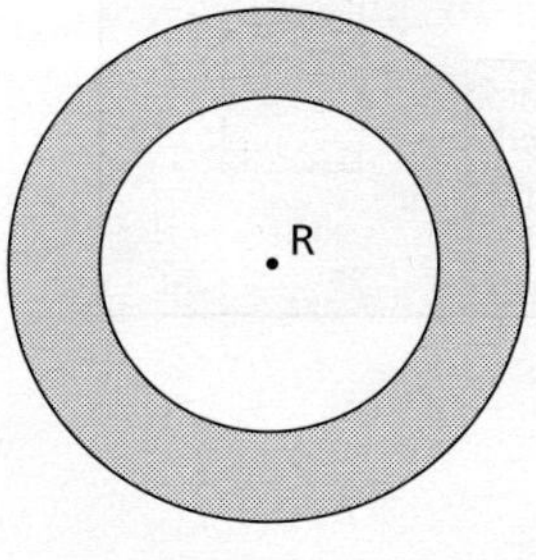

Point X can be marked anywhere in the shaded region.

A4

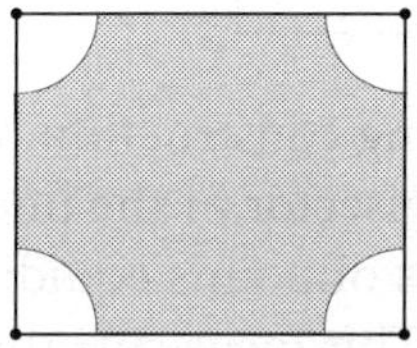

A5

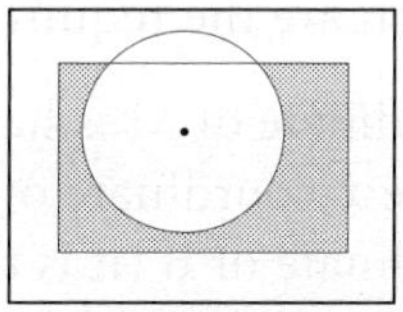

A6

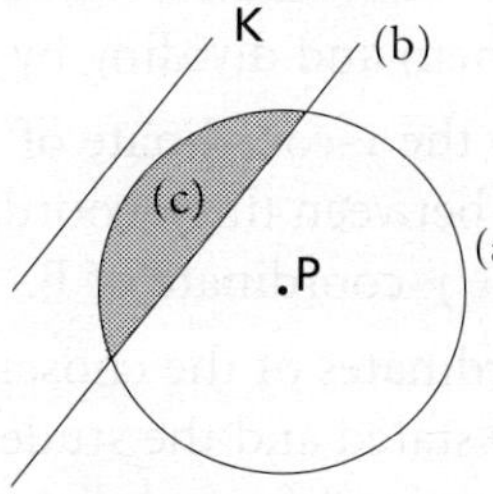

A7 An arrangement similar to this one

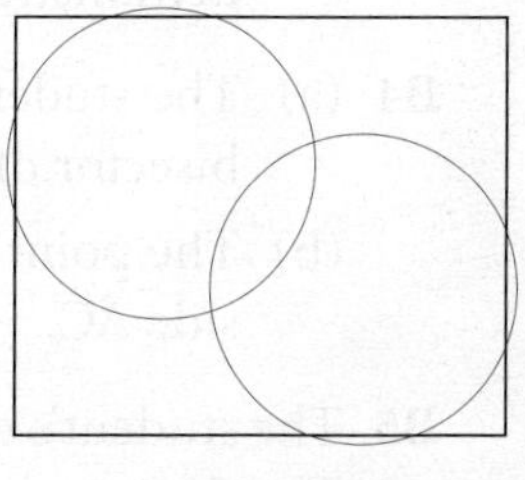

A8 (a) The set of points less than 2 cm from P.

(b) The set of points that are more than 2 cm from Q but are less than 3 cm from Q.

***A9** (a) The set of points that are less than 2 cm from l and less than 3 cm from T.

(b) The set of points that are less than 4 cm from R and less than 4 cm from S.

***A10** A figure 8

B *Nearer, further, the same distance …* (p 38)

B1

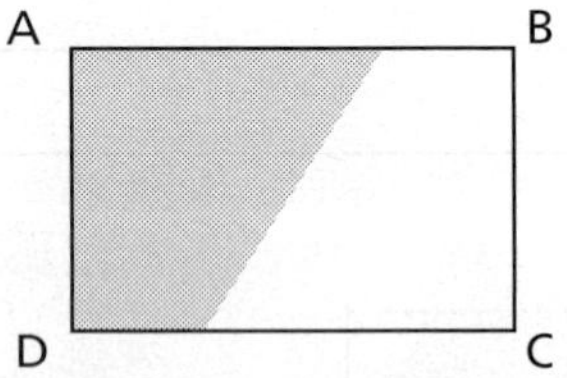

B2 Diagram showing construction of the perpendicular bisector of the line UV. This is the locus of points equidistant from U and V. This locus cuts the circle in two places, which are the required points.

B3 (a) The *x*-coordinate of M is half way between the *x*-coordinate of A and the *x*-coordinate of B (it is the mean of the two coordinates, found by adding them and dividing by 2).

Similarly the *y*-coordinate of M is half way between the *y*-coordinate of A and the *y*-coordinate of B.

(b) The coordinates of the chosen point on line *p* stated and the student's demonstration of equal distance.

B4 (a) The student's own triangle with the bisector of angle ABC drawn.

(b) The point where this line cuts side AC.

B5 The student's drawing.
The three loci should meet at a point.

B6 The circle touches all three sides of the triangle. The point where the three loci meet is equidistant from all three sides. So the three perpendicular lines from this point to each of the three sides have the same length, the radius of the circle. Being at right angles to these lines, the sides are tangents to the circle.

C *Mixed questions* (p 39)

C1

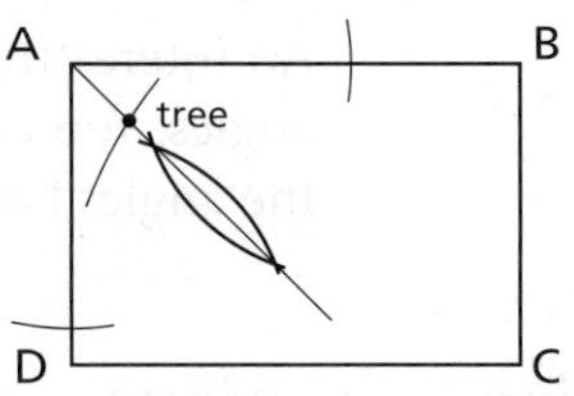

C2

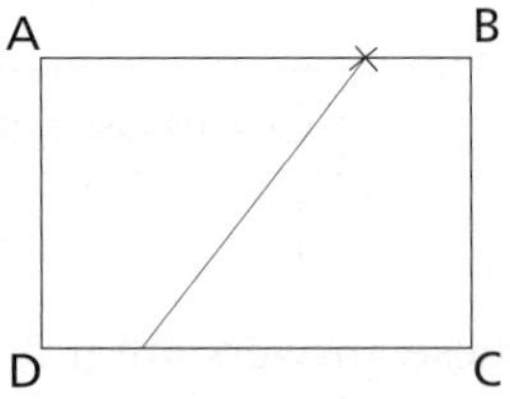

C3 (a) $28\,m^2$ (b) 28%

C4 (a) Yes (b) No

(c) Afzal, Cathy and Ebrahim

(d) Dave and Ebrahim

(e)

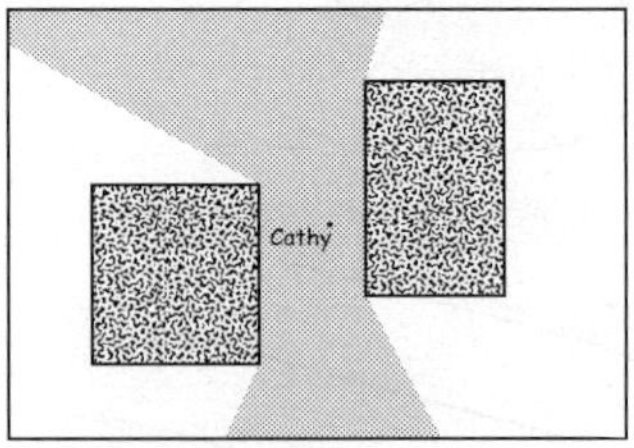

(f)

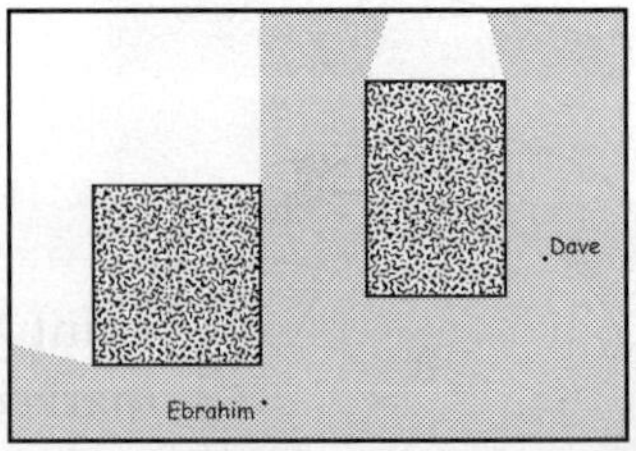

(g) It can be done (just) with two cameras like this.

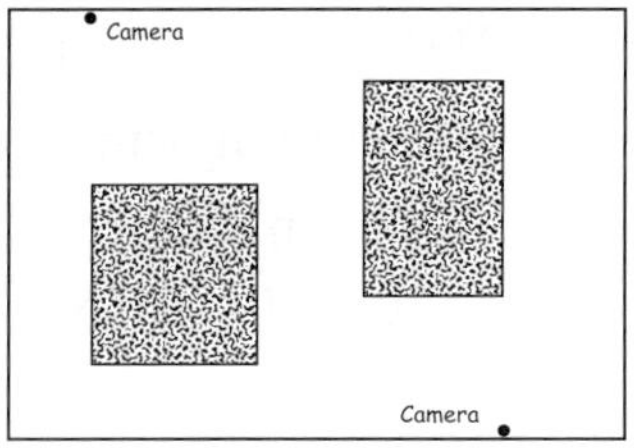

There are many possibilities with three cameras.

Test yourself (p 41)

T1

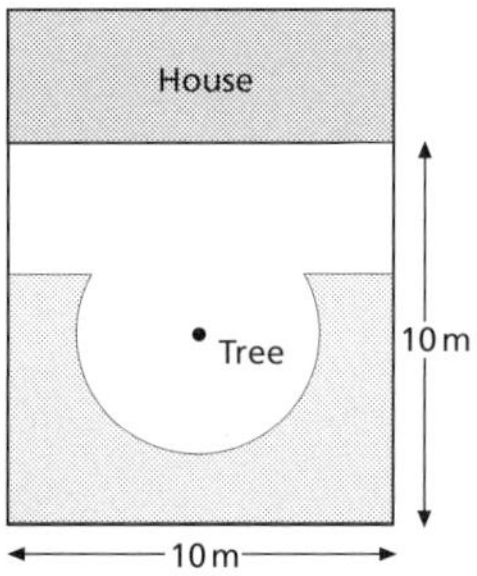

T2

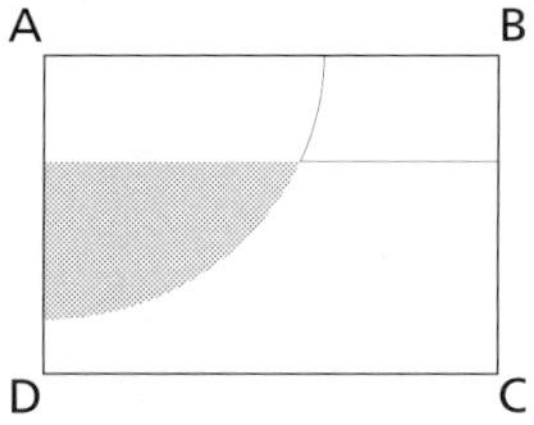

T3

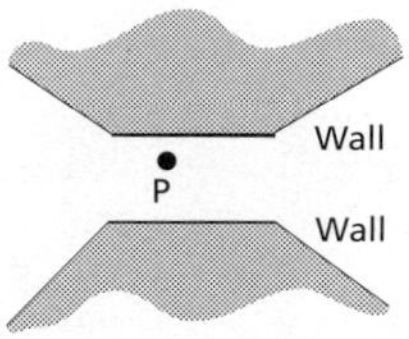

T4

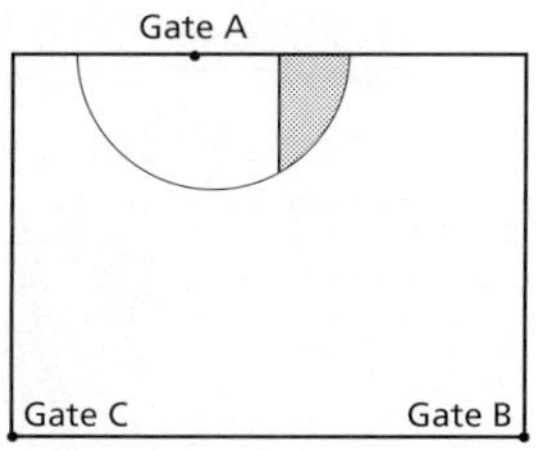

Practice booklet

Section A (p 6)

1

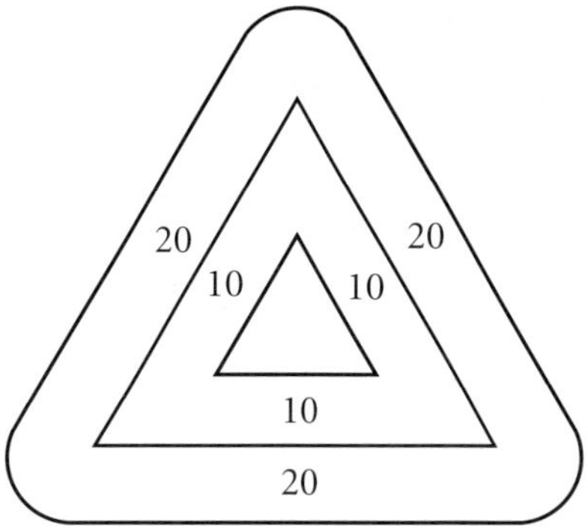

2 The set of points that are 3 cm or less from A, 3 cm or less from B and 2 cm or less from C.

3

4 (a) The set of points that are 3 cm or less from line *l* and 2 cm or less from line *m*.

(b) The set of points that are 3 cm or less from Q and 1 cm or more from P.

Sections B and C (p 7)

1

2

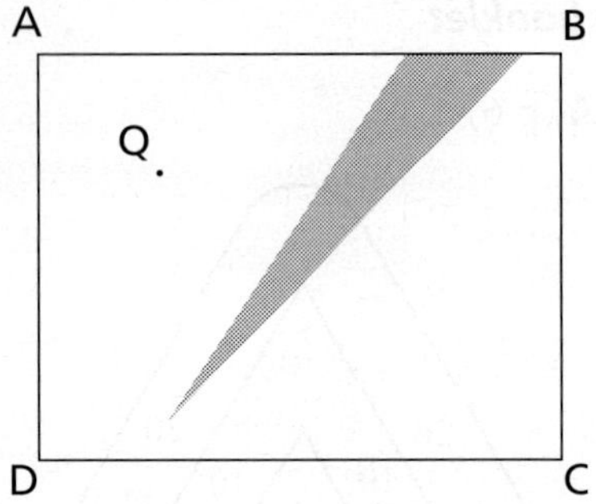

3

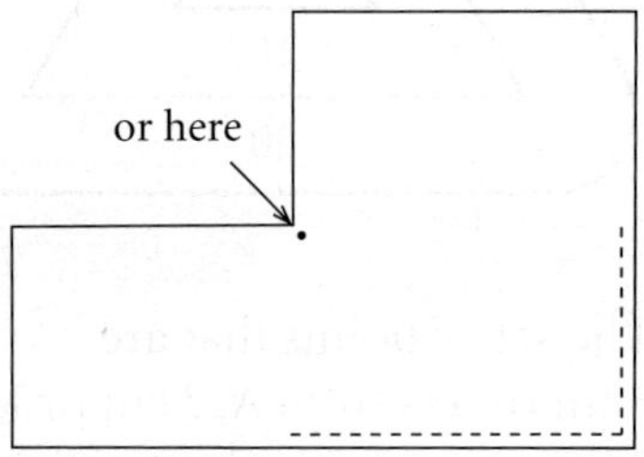

4 (a) The student's drawing

(b) The circle passes through the other two vertices

5 (a) The student's drawing

(b) The loci form diameters and meet at the centre of the circle.

5 Paired data

This unit revises KS3 work on drawing scatter diagrams and looking at the correlation between two variables. It extends the work to look at drawing lines of best fit.

Students will need to practice drawing scatter diagrams but much of the work in this unit could be dealt with using a computer spreadsheet or graphical calculator. The scatter diagrams for some later examples are reproduced on optional resource sheets to save excessive drawing of graphs.

Essential	**Optional**
Sheets P91, P92, P94	Sheet P93, P95, P96
Practice booklet pages 8 to 9	

A Connections (p 42)

Students should be familiar with scatter diagrams and correlation and this section offers plenty of practice using real data.

Sheet P91

◊ This section could be done entirely with the aid of a spreadsheet or graphic calculator. Students may need reminding of positive and negative correlation and judging whether there is a reasonable correlation or not. The data file with sample graphs can be downloaded from the SMP website (www.smpmaths.org.uk/resources) in Excel format. The information used to compile this database was mainly gleaned from www.whatcar.com.

◊ The issue of sensible interpretation of correlation in context is dealt with in section C but this could be usefully discussed in these examples.

◊ Some issues regarding graph drawing may arise here such as choosing scales and using a jagged start to show that a scale does not start at zero.

Scatter diagrams should be scaled so that the graph is roughly square in shape to give a true picture of any correlation.

Most of the diagrams in this unit use dots for points. This has the advantage that where points are repeated circles can be drawn around the dots. Some people prefer to use a cross or a circle with a small dot inside to give a more precise place on the graph.

◊ This table gives the actual correlations for the database.
For this quantity of data any correlation above 0.43 or below −0.43 is statistically significant.

	Price	m.p.g.	Engine	Power	Speed	Boot	Tank	Length	Width	Height	Weight
Price	1										
m.p.g.	−0.8	1									
Engine	0.81	−0.90	1								
Power	0.96	−0.87	0.90	1							
Speed	0.87	−0.84	0.90	0.96	1						
Boot	−0.23	0.04	0.01	−0.20	−0.15	1					
Tank	0.70	−0.83	0.81	0.72	0.72	0.22	1				
Length	0.39	−0.70	0.76	0.51	0.60	0.44	0.582	1			
Width	0.39	−0.57	0.51	0.40	0.42	0.23	0.64	0.63	1		
Height	−0.33	−0.02	−0.08	−0.29	−0.31	0.49	0.12	0.15	0.11	1	
Weight	0.66	−0.74	0.86	0.74	0.82	0.05	0.76	0.83	0.56	−0.31	1

B ***Lines of best fit*** (p 44)

Sheets P92, P94
Optional: Sheets P93, P95

◊ It is important to emphasise that it is only valid to draw a line of best fit when there is good positive or negative correlation. The examples in section A could be used to illustrate this point. Students should be encouraged to view estimation within the data range (interpolation) as reasonably accurate but outside the data range (extrapolation) as somewhat doubtful.

◊ Sheets P93 and P95 have the scatter diagrams for B2 and B4 should the students require no further practice in drawing scatter diagrams.

These crowded islands

Analysts often project population trends, usually with more sophisticated models than a linear one. This provides interesting discussion about how far one can reasonably extrapolate.

C ***Drawing conclusions*** (p 47)

This section looks at the interpretation of correlation and avoiding spurious conclusions. Question C2 is a case where the data clearly has a non-linear connection (its scatter diagram is on sheet P96).

Optional: Sheet P96

TG

◊ In explaining relationships students need to understand that correlation does not imply causality, it merely shows a numerical link. In many cases of apparently 'spurious' correlation there is a third variable to which the two being examined are linked. In this case the faster cars have larger engines and the structure to support them and hence will weigh more.

A likely story!

This is a well known case. Larger villages will of course have more chimneys and hence more storks' nests and larger villages will also have more births per annum.

A ***Connections*** (p 42)

A1 (a)

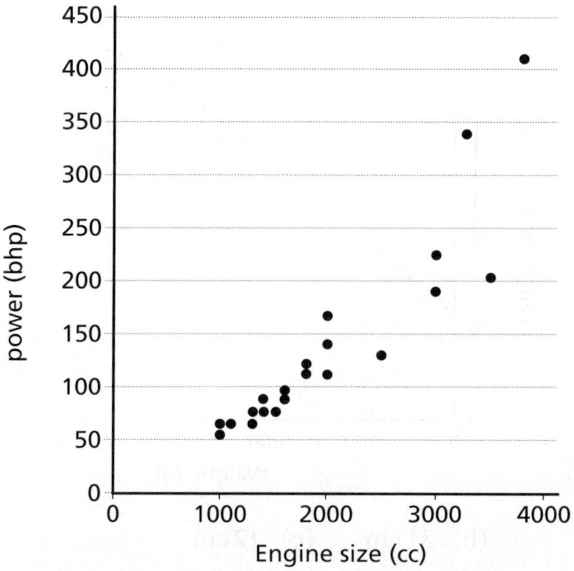

(b) Good positive correlation.

A2 (a)

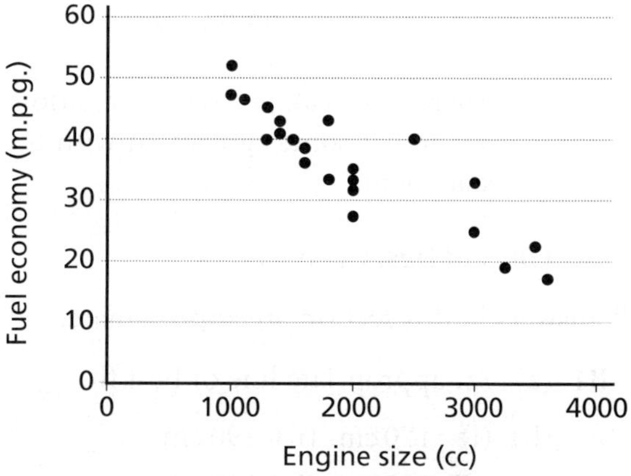

(b) Good negative correlation

(c) Larger engines use more petrol.

A3 (a)

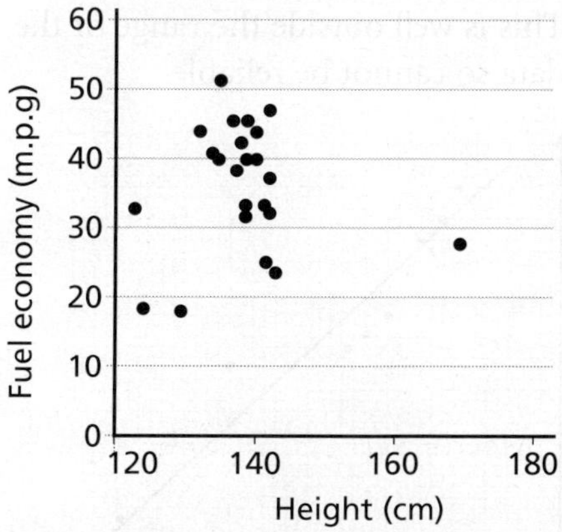

(b) There is no apparent connection between height and fuel economy.

A4 (a)

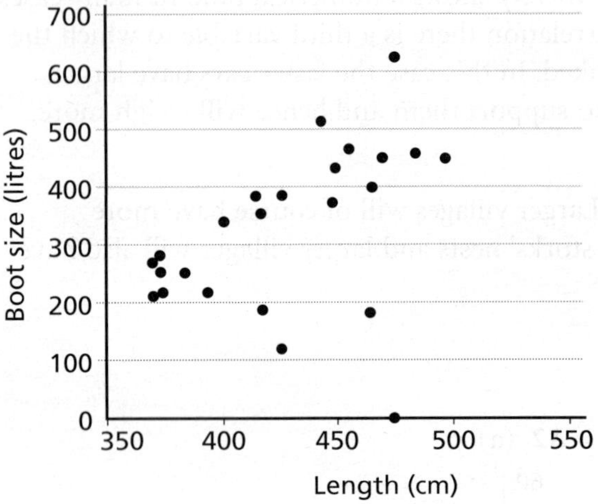

(b) There is a weak positive correlation so there is some agreement with the statement.

B *Lines of best fit* (p 44)

All answers in this section are approximate.

B1 (a) An appropriate line of best fit.

(b) (i) 170 cm (ii) 190 cm
(iii) 200 cm (iv) 140 cm

(c) (i) 54 cm (ii) 40 cm
(iii) 60 cm (iv) 27 cm

(d) 215 cm
This is well outside the range of the data so cannot be reliable.

B2

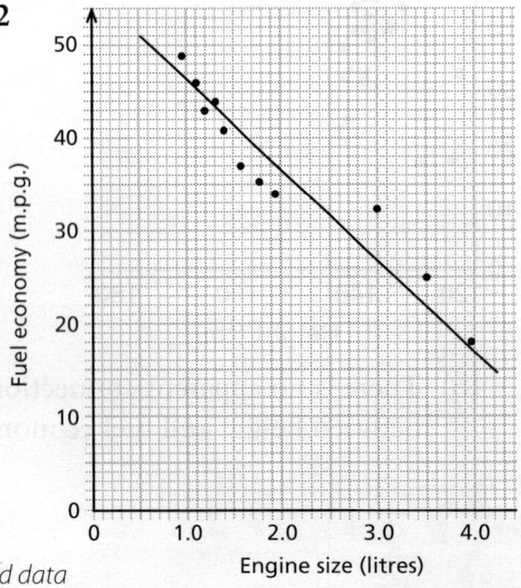

(c) Good negative correlation

(d) 32 mpg

(e) 0.7 litres

(f) 0 m.p.g. according to the graph. This is very unreliable!

B3 (a) Good positive correlation.

(b) An appropriate line of best fit, with roughly an even number of points each side of the line.

(c) Around 1800

B4 (a)

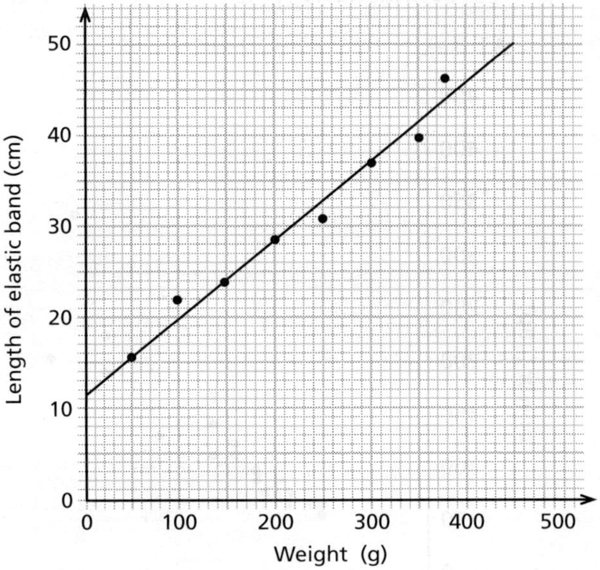

(b) 31 cm (c) 12 cm

(d) 50 cm

(e) No - the elastic band may even break before 1 kg is added.

C *Drawing conclusions* (p 47)

C1 (a)

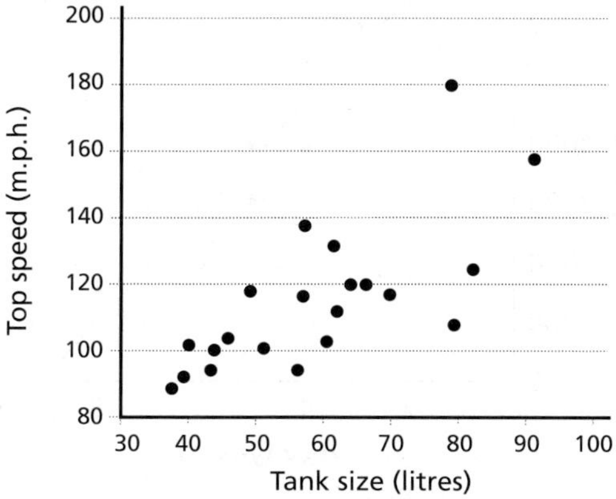

(b) There is a weak positive correlation. Cars which go faster generally have larger engines and lower m.p.g. so need larger tanks.

C2 (a)

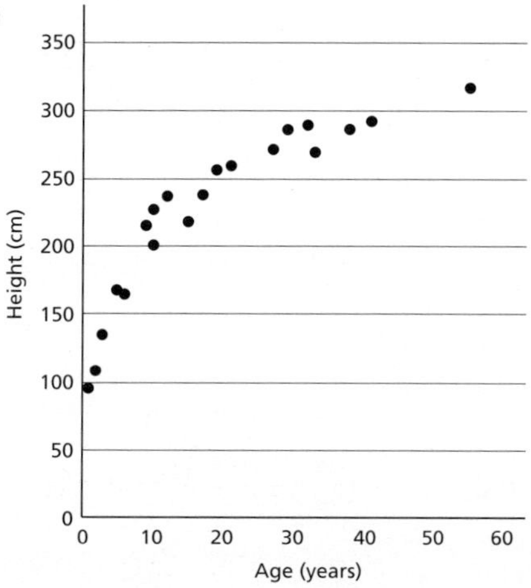

(b) Yes - about 300 cm

(c) The connection between height and age is not linear and it would be unreasonable to draw a straight line of best fit.

Test yourself (p 48)

T1 (a), (c)

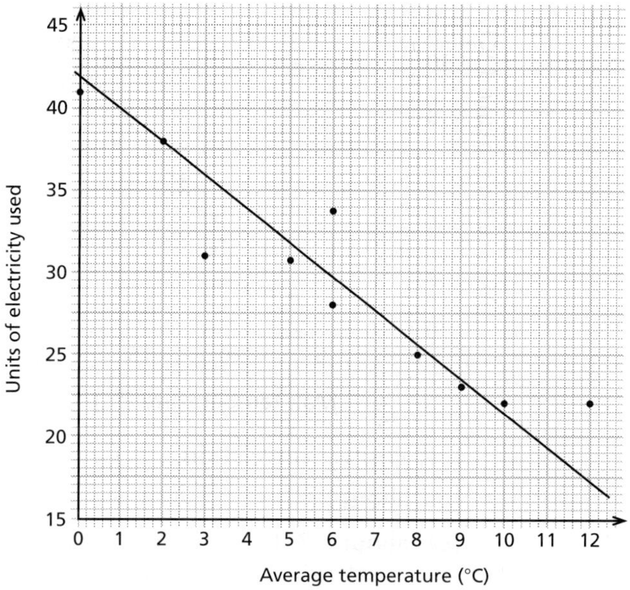

(b) A fairly good negative correlation.

(d) (i) 3.4°C (ii) 28 units.

Practice booklet

Section A (p 8)

1 (a) Positive correlation

(b) Yes, with student's reasons.

2 No, with student's reasons.

Section B (p 9)

1 (a), (b)

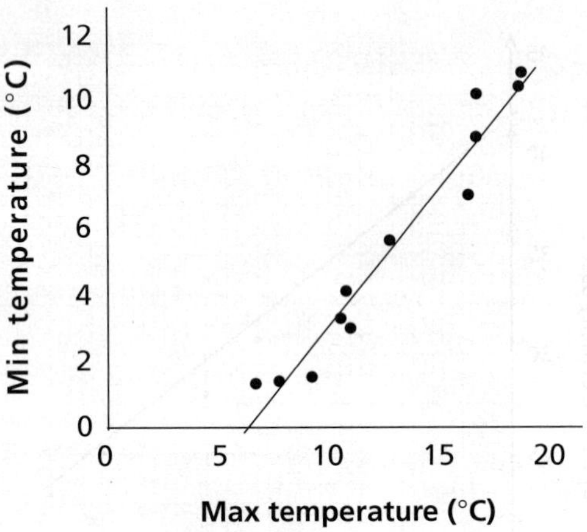

(c) Positive correlation.

(d) about 7 °C

(e) about 8.5 °C

2 (a)

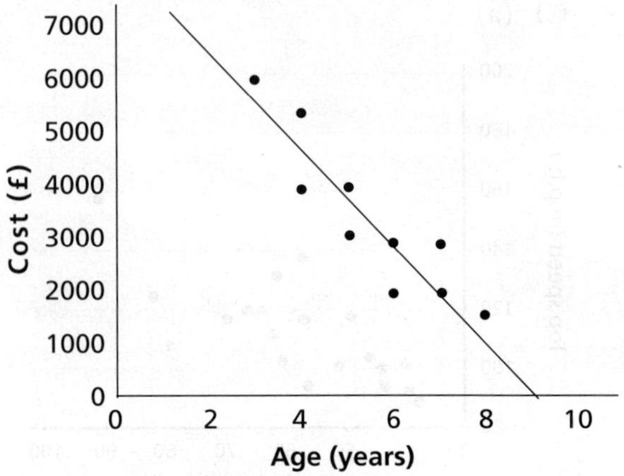

(b) about £6500

(c) No, this graph would give a negative price, and student's comments.

6 Forming and solving equations

Practice booklet pages 10 to 14

F *Further fractions* (p 49)

When discussing how to get rid of fractions, you may wish to mention that the smallest number we can multiply by is the Lowest Common Denominator (LCM), though of course any common denominator will serve just as well.

A *Review: unknown on one side* (p 49)

A1 (a) $x = 6$
(b) The student's check

A2 (a) $x = 4$ (b) $x = 22$
(c) $x = 32$ (d) $x = 35$

A3 (a) $n = 6$ (b) $n = 5$ (c) $n = \frac{1}{2}$

A4 (a) $x = 2\frac{1}{2}$ (b) $y = 4\frac{1}{2}$ (c) $z = 50$
(d) $n = {}^-3$ (e) $m = {}^-6$ (f) $p = {}^-3$

A5 (a) $x = 7$ (b) $y = 10$ (c) $x = 5$

A6 (a) $x = 0.4$ (b) $x = 0.5$ (c) $x = 20$
(d) $x = 2.6$ (e) $x = 0.1$ (f) $x = 0.8$

A7 (a) $f = 13$ (b) $e = 4$ (c) $f = 5$

B *Review: unknown on both sides* (p 50)

B1 (a) $x = 5$ (b) $x = 4$ (c) $x = 4$
(d) $x = 2$ (e) $x = 6$ (f) $x = 6$

B2 (a) $n = 5$ (b) $n = 2$ (c) $n = 3$
(d) $n = 2$ (e) $n = 3\frac{1}{2}$ (f) $n = 2$

B3 (a) $p = {}^-2$ (b) $k = {}^-\frac{1}{2}$ (c) $x = 4$
(d) $f = 0.4$ (e) $h = 0.4$ (f) $r = 0$

B4 (a) $d = 6$ (b) $d = 3$ (c) $d = 4$

B5 (a) $j = 3$ (b) $f = {}^-3$ (c) $h = 3$

C Equations from shapes (p 50)

C1 (a) $x = 40$, angles are 60°, 80° and 40°
(b) $x = 20$, angles are 35° and 55°

C2 (a) Sum of angles $= 9x$
(b) $x = 20$

C3 (a) $h = 6$, sides are 19, 25, 26 and 30
(b) $x = 8$, missing side is 30

C4 $x = 20$

C5 (a) (i) $10x + 4$
(ii) $2x(3x + 2) = 6x^2 + 4x$
(b) If $P = 44$, $x = 4$ and $A = 112$.

D Equations from words (p 52)

D1 (a) $n - 8$ (b) $2n - 8$
(c) $2n - 8 = 100$, $n = 54$
Alan has 54 DVDs and Beth has 46.

D2 (a) $y - 4$ (b) $2y$
(c) $4y - 4$
(d) $4y - 4 = 24$, $y = 7$
Chloe is 7, Becky is 3 and Emily is 14

D3 (a) $3x + 2 = 56$ (b) $x = 18$

D4 (a) $y + 5$ (b) $4y + 5$
(c) $4y + 5 = 61$, $y = 14$
The cost of one apple is 14p.

D5 (a) $2x + 21$
(b) $5x + 21 = 171$, $x = 30$

D6 (a) $2n + 12 = 3n$
(b) $n = 12$ and student's check

D7 Abel thinks of 29

D8 (a) $3n - 20 = 2n$, $n = 20$
(b) $2n + 3 = n + 10$, $n = 7$
(c) $\frac{1}{2}n + 14 = n$, $n = 28$

E Using brackets (p 54)

E1 (a) $n = 3$ (b) $m = 9$ (c) $x = 3\frac{1}{2}$
(d) $d = {}^{-}7\frac{1}{2}$ (e) $f = 17$ (f) $x = 3\frac{1}{2}$

E2 (a) $a = 4$ (b) $v = 2$ (c) $y = 2\frac{1}{2}$
(d) $x = 27$ (e) $u = 2$ (f) $r = {}^{-}11$

E3 $x = {}^{-}4$

E4 $x = 5\frac{1}{2}$

E5 (a) $a = 4$ (b) $x = \frac{1}{2}$

E6 (a) $a = 1\frac{1}{2}$ (b) $b = 4$ (c) $c = 5.6$

E7 $3(n - 5) = 21$, $n = 12$

E8 (a) $2(n + 7) = 40$, $n = 13$
(b) $3(n - 9) = 2n$, $n = 27$
(c) $5(n - 6) = n + 2$, $n = 8$

E9 $3(x - 4) + 2(3x + 2) = 19$
$3x - 12 + 6x + \mathbf{4} = 19$
$9x - \mathbf{8} = 19$
$9x = \mathbf{27}$
$x = \mathbf{3}$

E10 (a) $n = 2$ (b) $m = 5$
(c) $k = {}^{-}1$ (d) $h = 2$

***E11** $2(3x + 2) - 2(2x - 1) = 8$
$(6x + 4) - (4x - 2) = 8$
$6x + 4 - 4x + 2 = 8$
$2x + 6 = 8$
$2x = 2$
$x = 1$

***E12** (a) $x = {}^{-}3$ (b) $x = 2\frac{1}{2}$

F Further fractions (p 55)

F1 (a) $x = 5$ (b) $x = \frac{1}{4}$
(c) $x = {}^{-}5$ (d) $x = 4$
(e) $x = {}^{-}3$ (f) $x = 6$

F2 (a) $x = 17$ (b) $x = 1.2$

F3 (a) $\dfrac{2x-7}{3} + \dfrac{x-2}{4} = \dfrac{5}{6}$

$$\frac{12(2x-7)}{3} + \frac{12(x-2)}{4} = \frac{12 \times 5}{6}$$

$$\mathbf{4(2x-7) + 3(x-2) = 2 \times 5}$$

$$\mathbf{8x - 28 + 3x - 6 = 10}$$

$$\mathbf{11x = 44}$$

$$\mathbf{x = 11}$$

F4 (a) $x = 5$ (b) $x = 2$ (c) $x = 7$

F5 $x = 3\frac{1}{2}$

Test yourself (p 56)

T1 (a) $a = 28$ (b) $b = 7$ (c) $c = {}^{-}6$
(d) $d = 4\frac{1}{2}$ (e) $e = 0.6$ (f) $f = {}^{-}1$

T2 (a) $n = 5$ (b) $g = 4$ (c) $t = 4$
(d) $u = 2$ (e) $s = 9$ (f) $e = 6$

T3 (a) $x = 3$ (b) $x = 5\frac{1}{2}$

T4 (a) $x = 12$, angles are 72°, 60° and 48°
(b) $x = 37\frac{1}{2}$, angles are $52\frac{1}{2}$°, $52\frac{1}{2}$° and 75°

T5 (a) $(4x + 20)$°
(b) (i) $4x + 20 = 180$
(ii) $x = 40$, the smallest angle is 44°

T6 $x + 2(x + 4) = 24.5$, $x = 5.5$

T7 (a) $500x$
(b) $500x + 2 \times 25 = 1250$, $x = 2.4$

T8 (a) $a = 6$ (b) $b = 11$ (c) $g = 4$
(d) $x = {}^{-}27$ (e) $c = 3$ (f) $y = \frac{1}{2}$

T9 (a) $n = 6$ (b) $n = {}^{-}3$ (c) $n = 5$
(d) $n = 2\frac{1}{2}$ (e) $n = 3$ (f) $n = 1$

T10 (a) $x = 6\frac{1}{2}$ (b) $x = 12$ (c) $x = 3$

T11 (a) $2t + 4s$ (b) $4m + 12$ (c) $q = {}^{-}4$
(d) $a = {}^{-}0.8$ (e) $h = {}^{-}\frac{11}{7}$

Practice booklet

Section A (p 10)

1 (a) $x = 6$ (b) $x = 4$ (c) $x = 36$
(d) $x = 18$ (e) $x = 48$ (f) $x = 32$

2 (a)
$10 - 4x = 2$ [+ 4x]
$10 = 2 + \mathbf{4x}$ [– 2]
$\mathbf{8} = 4x$ [÷ 4]
$\mathbf{2} = x$

(b)
$7 = 13 - x$ [+ x]
$7 + \mathbf{x} = 13$ [– 7]
$x = \mathbf{6}$

(c)
$\frac{1}{3}x + 3 = 10$ [– 3]
$\frac{1}{3}x = \mathbf{7}$ [× 3]
$x = \mathbf{21}$

(d)
$\dfrac{2c-3}{5} = 3$ [+ 5]
$2c - 3 = \mathbf{15}$ [+ 3]
$2c = \mathbf{18}$ [÷ 2]
$c = \mathbf{9}$

3 (a) $x = 10$ (b) $c = 3\frac{1}{2}$ (c) $b = 24$
(d) $x = 4$ (e) $x = 2$ (f) $x = 160$

4 (a) $x = 16$ (b) $x = 5$ (c) $x = 7$

5 (a) $x = 2\frac{1}{2}$ (b) $x = 4$ (c) $x = 2.4$

Section B (p 11)

1 $x = 6$

2 (a) $3x + 20 = 6x + 2$
$x = 6$
(b) $2x + 132 = 6x + 12$
$30 = x$

3 (a) $x = 4$ (b) $x = 2$ (c) $x = 8$
(d) $x = 3$ (e) $x = 6$ (f) $x = 3$

4 (a) $5x + 18 = 8 - 5x$ $[+ 5x]$
$\mathbf{10x} + 18 = 8$ $[- 18]$
$\mathbf{10x} = {}^-10$ $[\div 10]$
$x = \mathbf{{}^-1}$

(b) $\frac{3}{4}p + 2 = 10 - \frac{1}{4}p$ $[+ \frac{1}{4}p]$
$\mathbf{p} + 2 = 10$ $[- 2]$
$\mathbf{p = 8}$

(c) $\frac{2z+3}{3} = 10$ $[\times 3]$
$2z + 3 = \mathbf{30}$ $[- 3]$
$2z = \mathbf{27}$ $[\div 2]$
$z = \mathbf{13\frac{1}{2}}$

5 (a) $x = 2.5$ (b) $x = 3.5$ (c) $x = 16$
(d) $x = 2$ (e) $x = 4$ (f) $x = 4$

Section C (p 12)

1 (a) $3a + 30$ (b) $3a + 30 = 180$
$a = 50$

(c) 50°, 60° and 70°

2 (a) A = 60° B = 100° C = 20°
(b) D = 90 E = 36° F = 54°
(c) G = 40° H = 75° I = 65°

3 (a) $4x + 1 = 15 - 3x$
$x = 2$

(b) length of rectangle is 9

4 (a) $3x + 30$ (b) $3x + 30 = 60$
$x = 10$

(c) Lengths are 10, 24 and 26 cm

5 (a) $6x + 6$ and $8x - 10$
(b) $6x + 6 = 8x - 10$
$x = 8$ cm

Section D (p 13)

1 (a) Pile B has $2x$; pile C has $x - 6$
(b) Total number of stones is $4x - 6$
(c) $4x - 6 = 94$
$x = 25$ (no. of stones in Pile A)

2 (a) (i) $6t$ (ii) $3(t + 5)$
(b) $6t + 3t + 15 + 3t = 207$
$12t + 15 = 207$
$t = 16$

A pencil costs 16p, a rubber 21p and a sharpener 48p

3 $b = 12$

4 Sam eats 61 bananas

Sections E and F (p 14)

1 (a) $3(s - 4) = 18$ [multiply out brackets]
$\mathbf{3s} - 12 = 18$ $[+ 12]$
$\mathbf{3s} = 30$ $[\div \mathbf{3}]$
$s = \mathbf{10}$

(b) L.H.S = $3(10 - 4) = 3 \times 6 = 18$ = R.H.S

2 (a) $x = 7$ (b) $y = 2$ (c) $z = 11$
(d) $x = 3$ (e) $d = 2$ (f) $x = 3$

3 (a) $n = 8$ (b) $p = 6$
(c) $t = 2$ (d) $p = 4$

4 (a) number is $10\frac{1}{2}$
(b) number is ${}^-6\frac{1}{2}$

5 $\frac{x+3}{4} = 2$ $[\times 4]$
$\frac{4(x+3)}{4} = 4 \times 2$ [cancel fraction]
$\mathbf{x + 3 = 8}$ $[- \mathbf{3}]$
$\mathbf{x = 5}$

6 $\frac{x+4}{3} = \frac{3x-2}{2}$ $[\times \mathbf{6}]$
$\frac{6(x+4)}{3} = \frac{6(3x-2)}{2}$ [cancel fractions]
$\mathbf{2(x + 4) = 3(3x - 2)}$ [**multiply out brackets**]
$\mathbf{2x + 8 = 9x - 6}$ $[- \mathbf{2x}]$
$\mathbf{8 = 7x - 6}$ $[+ \mathbf{6}]$
$\mathbf{14 = 7x}$
$\mathbf{x = 2}$

7 (a) $x = 14$ (b) $x = 11$

Review 1 (Student's book p 58)

1 (a) $n \le 3$ (b) $n > {}^-1$
(c) $1 \le n \le 5$ (d) $0 < n \le 4$

2 (a) $x = 5$ (b) $n = {}^-2$ (c) $x = 1.5$

3

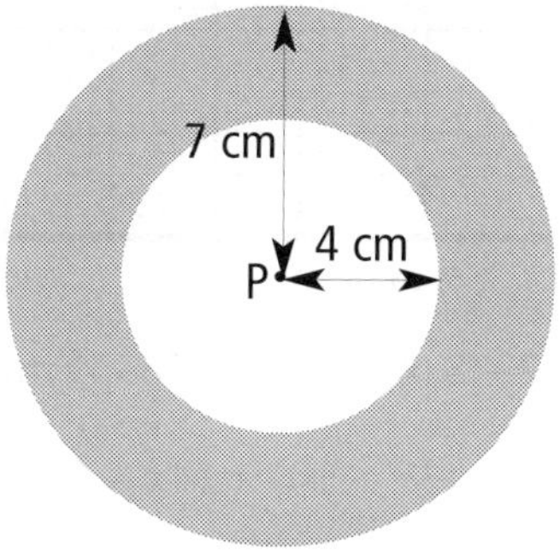

4 (a) $\frac{1}{2} \times \frac{1}{4}$ (b) $\frac{2}{3} \times \frac{4}{5}$ (c) $\frac{1}{4} \times \frac{4}{5}$
(d) $\frac{2}{3} \times \frac{1}{4}$ (e) $\frac{2}{3} \times \frac{3}{4}$

5 $c = 7$

6 (a)

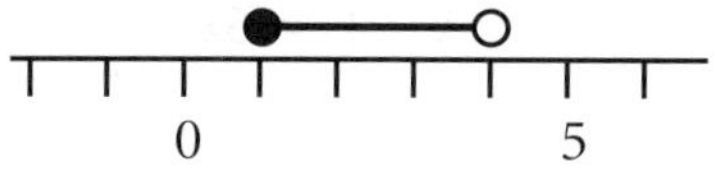

(b) $\sqrt{5}$, 1.9, π, $\frac{7}{12}$, 1

7 (a) 1
(b) Each is the reciprocal of the other.

8 (a) $4x + 16 = 28$
$x = 3$ so the width is 3 cm.
(b) $33\,\text{cm}^2$

9 (a) $10\frac{1}{2}$ tins (b) 16 days

10 (a) 10 (b) 25 (c) 2
(d) 14 (e) $\frac{4}{5}$

11 (a) $n + 3$
(b) $10n + 15(n + 3) = 95$ which simplifies to $25n + 45 = 95$ giving $n = 2$.
So the weight of a small marble is 2 grams.

12 $3 \le h < 4\frac{1}{4}$

13

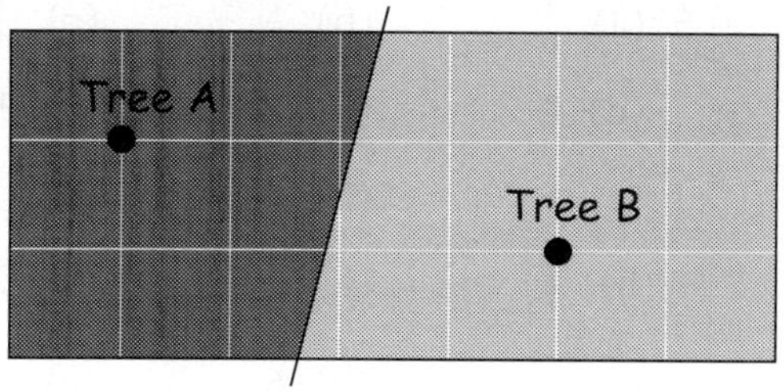

14 (a) and (c)

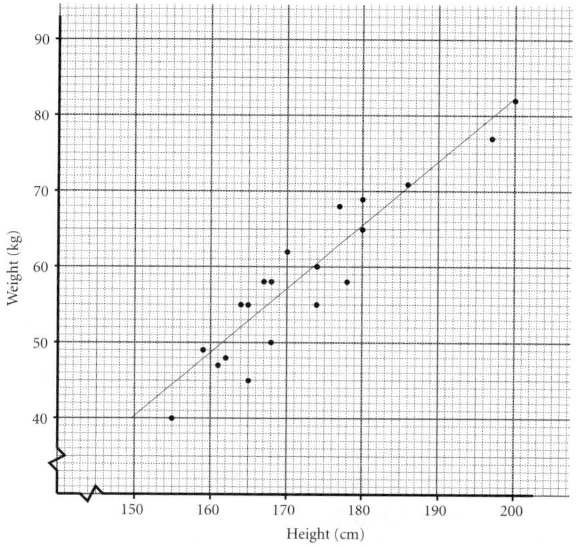

(b) The correlation is fairly strongly positive.
(d) 74 kg

15 $n = \frac{1}{2}$

Mixed questions 1 (p 15)

1 (a) T (b) F (c) T (d) F
(e) T (f) F (g) F (h) T

2 (a)

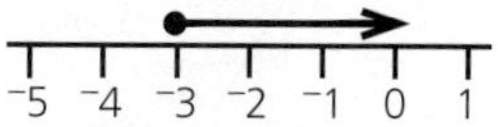

(b)

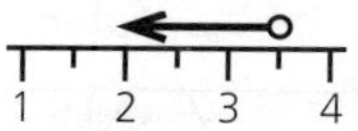

(c)

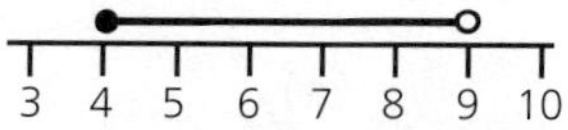

(d)

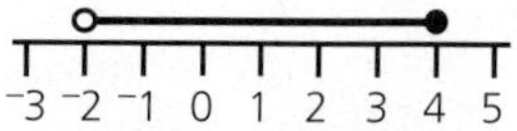

3 (a) ⁻1, 0, 1 (b) 49, 64 (c) 11, 13, 17

4 (a) $\frac{1}{7}$ (b) $\frac{5}{3}$ (c) $\frac{3}{7}$ (d) 10

5 (a) $\frac{1}{6}$ (b) $\frac{1}{2}$ (c) $\frac{1}{4}$

(d) 4 (e) $3\frac{1}{2}$ (f) 12

(g) 6 (h) $\frac{5}{3} = 1\frac{2}{3}$ (i) $\frac{1}{3}$

(j) $\frac{15}{16}$ (k) $\frac{4}{3} = 1\frac{1}{3}$ (l) 2

6 (a) $\frac{3}{5}$ (b) 6 (c) $\frac{3}{20}$

(d) $\frac{4}{15}$ (e) $\frac{15}{4}$

7

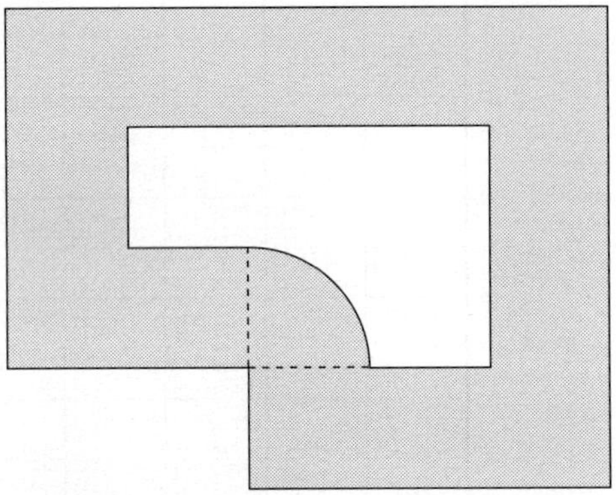

(To the required scale, the student's drawing should be 10 cm wide by 8 cm high.)

8

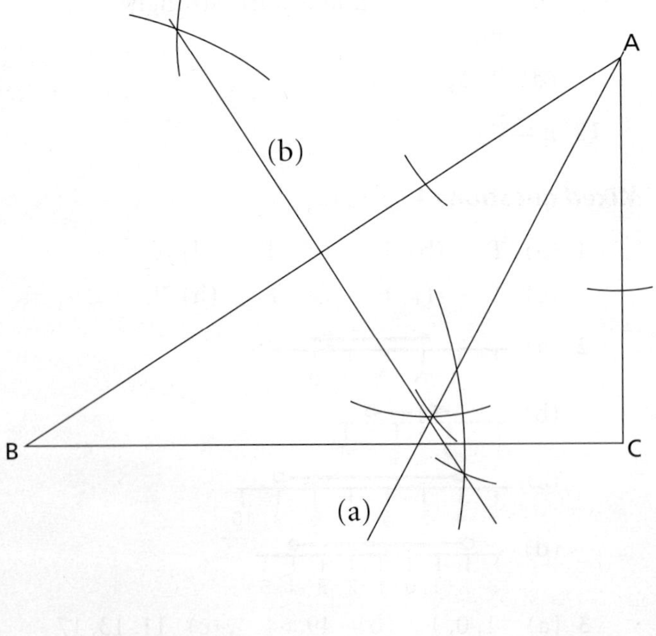

9

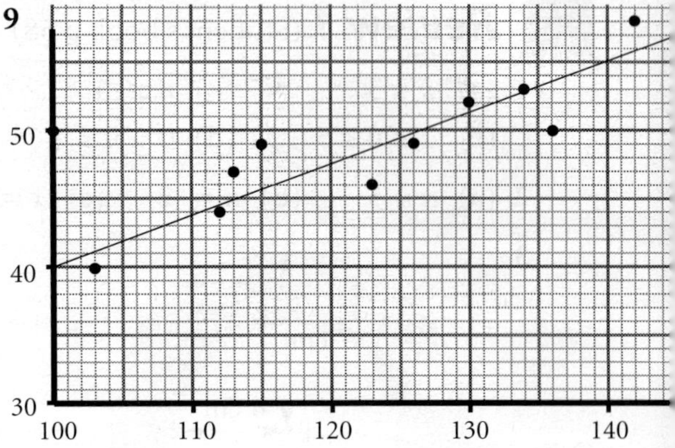

(c) 47–48 mm (d) 58–59 mm No

10 (a) $2x + (3x + 10) + (4x - 10) = 9x$

(b) $9x = 180° \Rightarrow x = 20°$ and angles are 40°, 70°, 70°

11 (a) (i) $4x$

(ii) $(x - 4) + (2x - 4) + (x - 4) + (2x - 4) = 6x - 16$

(b) $4x = 6x - 16$, $x = 8$
square: 8 cm by 8 cm
rectangle: 4 cm by 12 cm

12 (a) (i) $3x$ (ii) $x - 4$

(b) $(x - 4) + x + 3x = 56$
Ages 8, 12, 36

13 (a) $x = 7$ (b) $x = 10$ (c) $x = 4$

7 Large and small numbers

TG

TG

Essential	**Optional**
Scientific calculators	Rulers and 1p coins
Practice booklet pages 17 to 23	

A Powers of ten (p 60)

Scientific calculators
Students may need rulers and some 1p coins

TG

The two main objectives of the discussion are

- to define a million and a billion
- to revise powers of ten (including the rules for multiplying and dividing)

◊ Clarify the meaning of 1 million and 1 billion (1 million = 1 000 000 and 1 billion = 1 000 000 000). Discuss the fact that a billion used to be 1 million million, that is 1 000 000 000 000. In the 1970s, Parliament decided a billion was to be 1 000 000 000. France was the first to use 1 000 000 000 and USA 'imported' it in the 18th century. Parts of South America still use 1 000 000 000 000 as a billion.

◊ Students can try some of the problems at the top of the page (or invent their own).

◊ Discuss powers of ten in general, including multiplying and dividing any number by a power of ten. Revise the rules for multiplying and dividing powers.

D *Using a calculator* (p 64)

TG

The two main objectives of the discussion are to enable students to

- interpret numbers displayed in standard form on their calculators
- enter numbers in standard form in order to calculate with them

You should also discuss rounding to a certain number of significant figures or decimal places.

Common mistakes are

- pressing the '×' key in the middle of entering a number in standard form
- interpreting the display for, say, 4×10^{18} as 4^{18}.

E *Negative powers* (p 66)

E6 This revises multiplying and dividing powers of ten where the index can be negative. You may wish to postpone this until students calculate without a calculator in Section I.

A *Powers of ten* (p 60)

A1 (a) 10^3 (b) 10^5

A2 (a) 100 (b) 100 000
(c) 10 000 (d) 100 000 000

A3 (a) 9 (b) 14

A4 A and K B and L
C and F D and E
G and J H and I

A5 (a) 10^5 (b) 10^9 (c) 10^3 (d) 10^6

A6 A and F B and G
C and D E and H

A7 (a) 60 000 000 (b) 230 000
(c) 5 000 000 (d) 13 000 000
(e) 20 700 000 000 (f) 400 000

A8 (a) 120 (b) 51 600 000
(c) 20 900 (d) 63 000
(e) 294 000 000 (f) 7 620 000 000

B *Writing large numbers* (p 61)

B1 C (3.13 thousand km)

B2 (a)

Yangtze-Kiang	6300 km
Zaire	4700 km
Danube	2840 km
Seine	780 km

(b) 6

B3 100 000 or 10^5 or a hundred thousand

B4 (a)

Year	Population	Billions
1900	1 630 000 000	1.63
1950	2 520 000 000	2.52
1960	3 000 000 000	3
1970	3 700 000 000	3.7
1980	4 500 000 000	4.5
1990	5 300 000 000	5.3
1995	5 690 000 000	5.69

(b) 2

B5 C 'Desert is about $\frac{1}{4}$ of the total land area of Earth.'

B6 (a) The average price is
£4.2 billion ÷ 230 billion
= £0.018260869 …
= £0.02 (to the nearest penny)
= 2p

An alternative method is to say that the average price is about
£4 billion ÷ 200 billion
= £4/200
= 2p

(b) The total amount spent on biscuits by one household in a year is £3.50 × 52 = £182.
So the number of homes is
£4.2 billion ÷ £182
= 4 200 000 000 ÷ 182
= 23076923.08 …
= 20 000 000 (to one s. f.)
= 20 million

An alternative method is to say that the number of homes is about
£4 billion ÷ £200
= 4 000 000 000 ÷ 200
= 20 000 000
= 20 million

C *Standard form* (p 63)

C1 A and D (1.8×10^9 and 6×10^5)

C2 (a) 30 000 (b) 20 000 000
(c) 800 000 000 000
(d) 420 000 (e) 810 000 000
(f) 5 300 000 (g) 937
(h) 7 040 000 (i) 5 890 000 000

C3 (a) 9×10^4 (b) 5×10^7
(c) 8×10^{11} (d) 2×10^{15}
(e) 6×10^6 (f) 3×10^9

C4 (a) 6.3×10^4 (b) 9.1×10^9
(c) 7.6×10^{13} (d) 3.29×10^8
(e) 5.73×10^{17} (f) 1.278×10^{14}

C5 3.84×10^5 km

C6 9 460 000 000 000 km

C7 (a)

Chinese	1 093 000 000
English	450 000 000
Hindi	367 000 000
Spanish	352 000 000
Russian	204 000 000

(b) The number of Mandarin speakers was actually about 5 times the number of Russian speakers.

C8 (a)

Baha'ism	5.8×10^6
Buddhism	3.4×10^8
Christianity	1.9×10^9
Confucianism	6.3×10^7
Hinduism	7.6×10^9
Islam	1.0×10^9
Jainism	4.0×10^6
Judaism	1.4×10^8
Shintoism	3.4×10^6
Sikhism	2.0×10^7

(b) Hinduism, Christianity, Islam, Buddhism, Judaism, Confucianism, Sikhism, Baha'ism, Jainism, Shintoism

C9 4×10^5

D Using a calculator (p 64)

D1 (a) 6×10^{13} (b) 1.6×10^{13}
(c) 1.24×10^{10} (d) 4.925×10^{15}

D2 (a) 3.83×10^{15} (b) 3.44×10^{10}

D3 (a) 2.46×10^{7} (b) 2.171×10^{15}
(c) 5×10^{10} (d) 5.4×10^{6}

D4 (a) 9.6×10^{11} (b) 2.7×10^{12}
(c) 6.5×10^{13} (d) 4.2×10^{18}

D5 (a) 52 000 000 000 (b) 3 100 000
(c) 22 000 000 000

D6 (a) 3.15×10^{9} (b) 1.955×10^{18}
(c) 8.85×10^{9}

D7 4.5×10^{6}

D8 2.97×10^{11}

D9 19984

D10 5.8×10^{9} kg

D11 20 000 000 kg

E Negative powers (p 66)

E1 A and H B and I C and F
D and J E and G

E2 (a) $\frac{1}{10}$ (b) $\frac{1}{1000}$ (c) $\frac{1}{100000}$

E3 (a) 0.0001 (b) 0.00001
(c) 0.1 (d) 0.000001

E4 (a) 60 (b) 4100 (c) 59
(d) 5.6 (e) 88.2 (f) 0.451
(g) 0.39 (h) 0.467 (i) 0.395
(j) 0.063 (k) 0.000 03
(l) 0.000 000 07

E5 (a) $a = {}^{-}6$ (b) $b = {}^{-}7$

E6 (a) 10^{2} (b) 10^{-3} (c) 10^{1} (d) 10^{-5}
(e) 10^{-2} (f) 10^{-1} (g) 10^{-6} (h) 10^{5}

F Writing small numbers (p 67)

F1 (a) 0.0091 (b) 0.0621
(c) 0.000 34 (d) 0.000 053
(e) 0.000 000 192 (f) 0.000 000 000 69

F2 (a) 0.02 (b) 0.0006
(c) 0.005 (d) 0.000 04
(e) 0.000 003 (f) 0.000 000 000 02

F3 (a) 2.9×10^{-3} (b) 4.8×10^{-5}
(c) 9.23×10^{-2} (d) 6.45×10^{-6}
(e) 1.9×10^{-9} (f) 8.2×10^{-11}
(g) 3×10^{-4} (h) 7×10^{-7}
(i) 9×10^{-12}

F4 1.4×10^{-5} kilograms

F5 0.00043 metres

F6 House mouse
Helena's hummingbird
Kitti's hog-nosed bat
Bee hummingbird
Pygmy shrew
House spider

G Small numbers on your calculator (p 68)

G1 (a) 1.156×10^{-5} (b) 2.1×10^{-7}
(c) 1.6×10^{-6}

G2 (a) 2.2×10^{-3} (b) 3.1×10^{-5}
(c) 3.7×10^{-4} (d) 2.6×10^{-11}
(e) 3.0×10^{-6} (f) 2.0×10^{-11}

G3 (a) 0.0311 (b) 0.00475

G4 8.5×10^{-5}

H Mixed problems (p 68)

H1 (a) (i) A $((9.46 \times 10^{12}) \times 2\,200\,000)$
(ii) 2.0812×10^{19} km

(b) (i) $\left(\frac{3.217 \times 10^{14}}{9.46 \times 10^{12}}\right)$
(ii) 34 light years

H2 (a) £3.604×10^{10} (b) £1.07×10^3

H3 26.1%

H4 (a) 9.1×10^{-25} (b) 4.55×10^{-18}

H5 (a) 5.98×10^{21} (b) 81
(c) 2.58×10^{11} (to 3 s. f.)

H6 (a) 2.79×10^{-6} cm^2
(b) 1.024×10^{-2} cm

H7 (a) 2.4×10^9 (b) 8.64×10^{12}

Calculating without a calculator (p 70)

I1 (a) 3×10^8 (b) 2.6×10^5
(c) 1.49×10^9 (d) 5×10^6

I2 (a) 8×10^{10} (b) 9×10^7
(c) 1.8×10^7 (d) 3.5×10^{10}

I3 (a) 4×10^2 (b) 4.5×10^7
(c) 5×10^1

I4 $(2 \times 10^4) + (5 \times 10^3)$ = 20 000 + **5000** = **25 000**

I5 (a) 240 000 (b) 30 000 (c) 20
(d) 0.5 (e) 25

I6 3.8×10^{20}

I7 (a) 5×10^3 (b) 3×10^1
(c) 6.3×10^{-4} (d) 5×10^{-5}

I8 (a) 6×10^3 (b) 4.8×10^{-5}
(c) 2×10^{-2} (d) 3.5×10^{-3}
(e) 1.3×10^{-4}

I9 1.2×10^{-18}

Test yourself (p 71)

T1 (a) 5.78×10^7
(b) About 236 people per square km

T2 140 000 000 or 1.4×10^8 years

T3 (a) 4.746×10^7
(b) 2.67×10^8 (to 2 s. f.)

T4 (a) 4.8 centimetres
(b) 5.8 grams

T5 (a) 5.236×10^{12} metres
(b) 9.454×10^{15} metres
(c) 5.538×10^{-4} light years

T6 (a) 25 300 000 (b) 0.000052

T7 (a) 1.6×10^8 (b) 28 000

Practice booklet

Section A (p 17)

1 a thousand

2 (a) 10^4 (b) 10^3 (c) 10^5 (d) 10^7

3 E F D H A G C J B

4 (a) 10^5 (b) 10^6 (c) 10^2 (d) 10^3

5 G A H D E B C F

6 (a) 30 000 (b) 17000 000
(c) 2 500 000 000 (d) 2 500 000
(e) 159 000 (f) 6 400 000

7 (a) $5.26 \times 10^4 = 526 \times 100$
$52.6 \times 10\,000 = 562 \times 10^3$
(b) $72.3 \times 10^5 = 7.23 \times 1\,000\,000$
$0.723 \times 1\,000\,000 = 723 \times 10^3$

8 (a) 2 (b) 4 (c) 1
(d) 8 (e) 4

Section B (p 18)

1 (a) 2 300 000 (b) 5 000 000 000
(c) 510 000 000 (d) 380 000
(e) 2 330 000 (f) 227 900 000

2 (a) 27 (b) Bus, taxi, tractor
(c) 31 575 000 (d) 32 176 000
(e) 601 000 (f) 8

3 (a) 10 000 000 000 (10 billion)
(b) 20 000 000 kg
(c) 1250 000 million

Section C (p 19)

1 (a) 2000 (b) 700 000
(c) 6 000 000 (d) 2 500
(e) 31 000 (f) 8 900 000
(g) 103 (h) 56 300
(i) 981 000

2 (a) 4×10^4 (b) 3×10^5 (c) 6×10^9
(d) 1×10^7 (e) 5×10^{12} (f) 7×10^{11}

3 (a) 2.6×10^4 (b) 6.2×10^8
(c) 1.49×10^{10} (d) 7.86×10^{12}
(e) 6.005×10^{13} (f) 2.149×10^{14}

4 (a) 6.1×10^4 (b) 2.5×10^7
(c) 4.5×10^9 (d) 3.7×10^5
(e) 9×10^3 (f) 7.2×10^4
(g) 5×10^5

5 (a) 4 (b) 4 (c) 3
(d) 6 (e) 2 (f) 1

6 (a) 7×10^3 (b) 6×10^4
(c) 4.5×10^6 (d) 5.07×10^4
(e) 9×10^7 (f) 7.2×10^5

Section D (p 19)

1 (a) 4×10^{11} (b) 6.4×10^9
(c) 2×10^{14} (d) 5×10^2
(e) 2×10^{10} (f) 2.50064×10^{13}

2 (a) 2.25×10^{11} (b) 9.17×10^6
(c) 1.20×10^{10} (d) 2.92×10^{11}
(e) 1.20×10^7 (f) 3.04×10^8

3 (a) 2.0×10^9 (b) 1.0×10^{15}
(c) 2.4×10^9 (d) 1.8×10^{12}
(e) 1.0×10^{11} (f) 7.9×10^7
(g) 6.9×10^5 (h) 9.0×10^2

4 (a) 4 000 000 (b) 2 500 000
(c) 54 000 000 (d) 28 000
(e) 38 (f) 1400

5 1.728×10^8

Sections E and F (p 20)

1 F C D B G A E

2

×	10^3	10^{-2}	10^5	10^{-7}
10^{-5}	10^{-2}	10^{-7}	10^0	10^{-12}
10^5	10^8	10^3	10^{10}	10^{-2}
10^{-1}	10^2	10^{-3}	10^4	10^{-8}
10^{10}	10^{13}	10^8	10^{15}	10^3

3 (a) 0.35 (b) 0.0268
(c) 0.000124 (d) 560
(e) 0.000 8 (f) 0.562

4 (a) 0.4 (b) 0.0008
(c) 0.0072 (d) 0.000 000 106
(e) 0.000 004 31 (f) 0.000 000 000 098

5 (a) 4×10^{-2} (b) 6.2×10^{-4}
(c) 7.51×10^{-6} (d) 4.27×10^{-5}
(e) 6.84×10^{-3} (f) 8.3×10^{-9}

6 5×10^{-9}

7 0.0025

Section G (p 21)

1 (a) 1.395×10^{-7} (b) 2×10^{-7}
(c) 4.096×10^{-9}

2 (a) 1.6×10^{-4} (b) 4.0×10^{-3}
(c) 1.7×10^4 (d) 6.4×10^{14}
(e) 7.6×10^{-7} (f) 1.9×10^9
(g) 2.9×10^{-2} (h) 1.3×10^5

3 (a) 9.4×10^{-5} (b) 3.17×10^{-10}
(c) 1.11×10^{-15} (d) 3.87×10^{-2}

4 1.02×10^{-3} or 1.0×10^{-3}

Section H (p 22)

1 (a) 1.5×10^2 (b) 2.4×10^7

2 5×10^8 litres

3 0.09 or 0.1 mm

4 (a) 1992 : 3.283×10^7
1997 : 3.049×10^7
2001 : 2.637×10^7

(b) 1992 : 43%
1997 : 31%
2001 : 32%

(c) 8.2×105

5 3.018×10^{-23}

Section I (p 23)

1 (a) 4.5×10^5 (b) 3.7×10^6
(c) 1.83×10^7 (d) 6.4×10^2

2 (a) 1.2×10^4 (b) 1×10^5
(c) 5.4×10^4 (d) 4.6×10^4
(e) 2×10^8 (f) 8×10^{-2}

3 (a) 6×10^9 (b) 4.2×10^{10}
(c) 4×10^6 (d) 4.9×10^{19}

4 (a) 2×10^2 (b) 5×10^4
(c) 1.6×10^2 (d) 2.3×10^4

5 (a) 3.4×10^{-1} (b) 5×10^{-2}
(c) 4.78×10^{-4} (d) 8.8×10^{-5}

6 (a) 2.1×10^3 (b) 1.8×10^{-7}
(c) 3×10^4 (d) 6×10^1

7 (a) 364 000 (b) 75 000
(c) 90 000 (d) 21
(e) 2.8 (f) 40 000

8 (a) 2×10^7 (b) 2.5×10^{13}

8 3D vision

This unit revises the different ways of representing 3 dimensional objects in 2 dimensions and reflection symmetry in 3D objects.

Essential	**Optional**
Sheets P97, P98, P99 Multilink cubes, mirrors Triangular spotty paper, centimetre squared paper	Soma Cube pieces

Practice booklet pages 24 to 26

A ***The Soma Cube*** (p 72)

Triangular spotty paper
Optional:
Multilink cubes, Soma Cube pieces

◊ The student interest in this section will be enhanced by allowing them to try to put together the Soma Cube. These can be bought from mathematical suppliers or made from multilink cubes. The set in the picture were made easily from a 25 mm square length of wood. There is a wealth of Internet resources on the Soma Cube including a downloadable interactive package to construct a cube. The term 'Soma' refers to the addictive drug described in Aldous Huxley's *Brave New World*, as the inventor of the puzzle found it addictive.

B ***Nets*** (p 73)

Centimetre squared paper, triangular spotty paper
Optional:
Scissors, compasses

◊ There are 12 distinct nets for a cube.

Looking for these as a group exercise is very valuable as it will involve discussion of reflection and rotation symmetry. A square cannot be joined to the net by a point only.

C *Views* (p 74)

Centimetre squared paper

◊ In conventional technical drawing the three views are arranged in a particular way. Here any correct view is accepted, regardless of orientation.

D *Reflection symmetry* (p 76)

Sheet P97 Multilink cubes - 2 different coloured sets of 5 per pair Mirrors, triangular spotty paper

Test yourself (p 78)

Sheets P98, P99

A *The Soma Cube* (p 72)

A1 D has 3, the rest have 4

A2 A, B, C, E, F, G

A3 Student's drawings of A, B, C, D, F, G

B *Nets* (p 73)

B1 Only A

B2 Four nets of a cube.
(See main text for list.)

B3 There are only two unique nets for a regular tetrahedron.

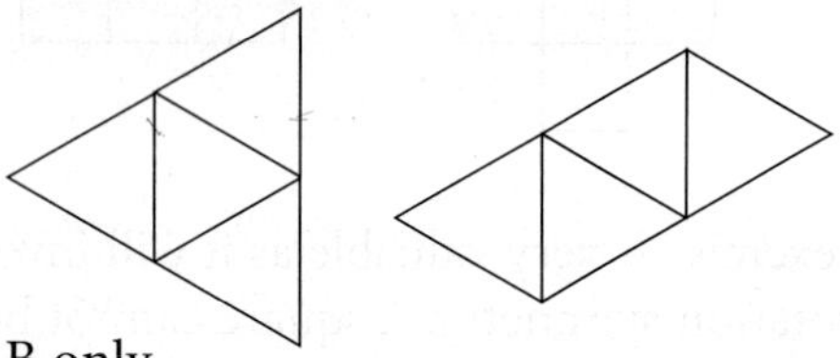

B4 B only

B5 (a)

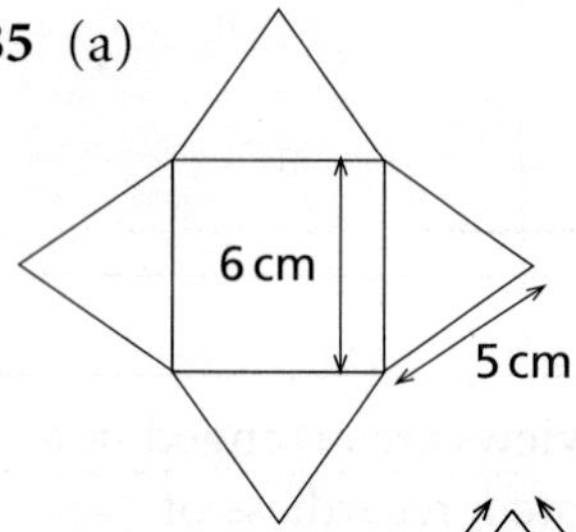

(b)

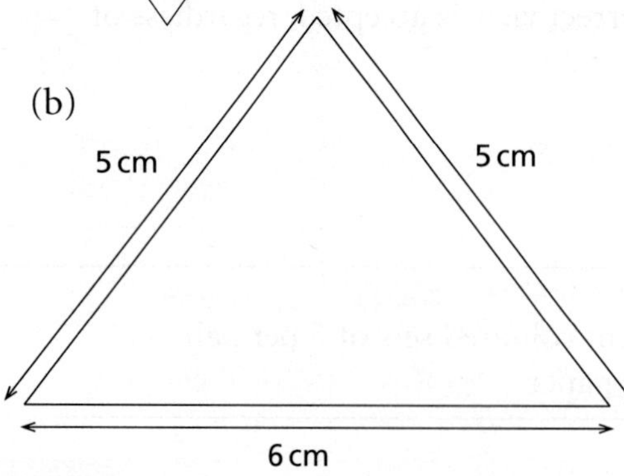

C *Views* (p 74)

C1 B or C

C2

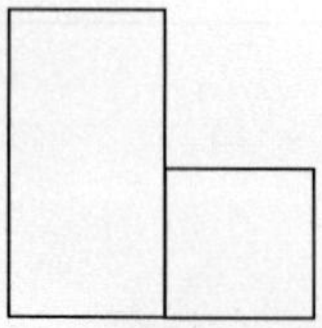

C3 B

C4 Plan, front and side view of another Soma Cube piece.

C5 Other orientations may be used.

(a)

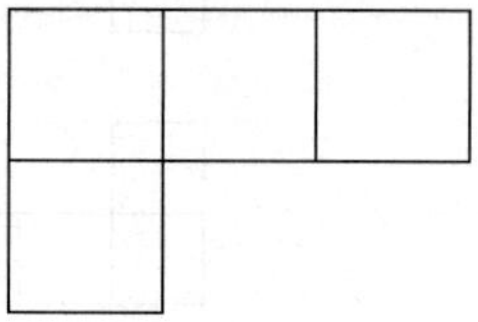

(b)

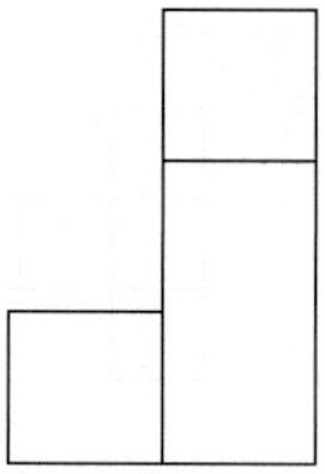

(c)

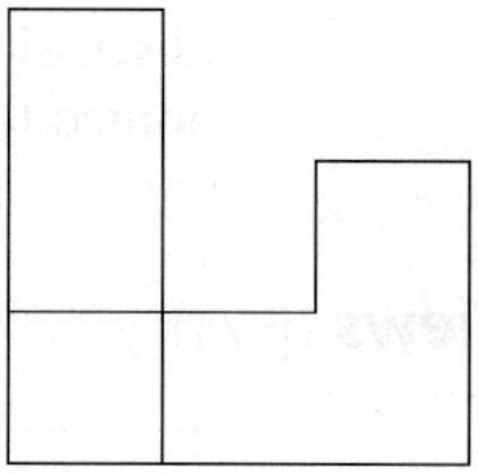

C6 (a)

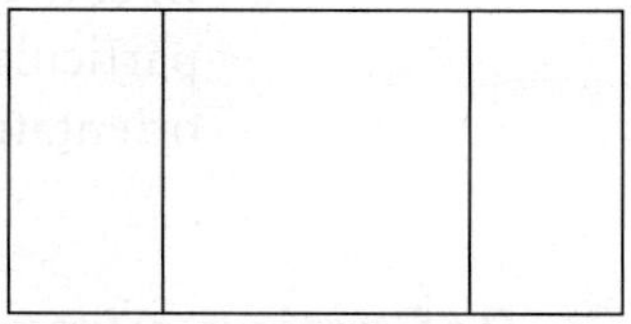

(b)

(c) 2.8 cm

C7 (a)

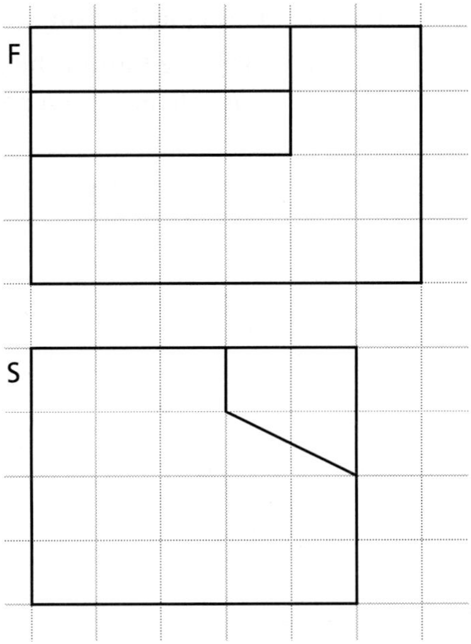

(b) 2.2 cm

D *Reflection symmetry* (p 76)

D1 D

D2 A and G, B and H, C and E, D and F

D3

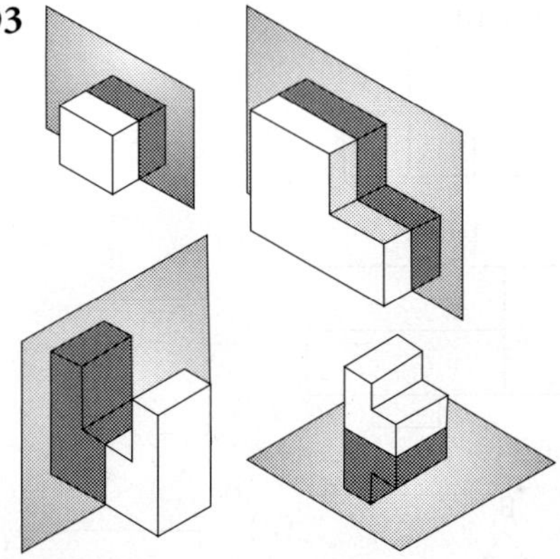

D4 (a) yes, 2 (b) No
(c) Yes, 3 (d) Yes, 1

D5

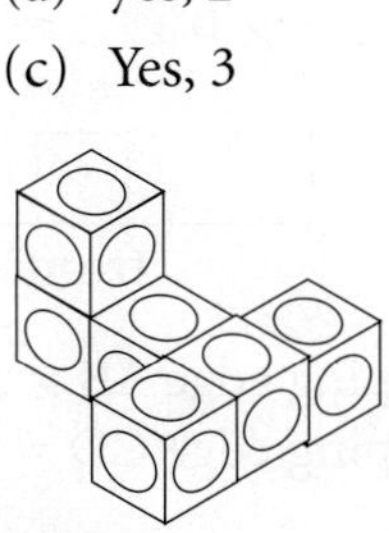

or similar

***D6** (a) 9 planes (b) 1 plane
(c) 5 planes (d) 5 planes

Test yourself (p 78)

T1 (a) $30\,\text{cm}^2$
(b)

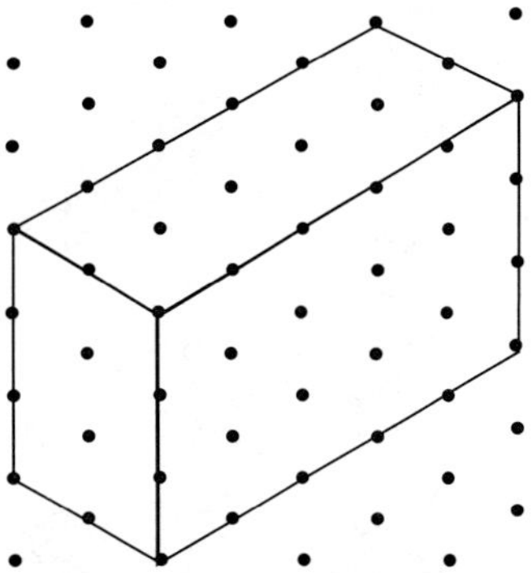

T2 (a) Student's full size net of prism
(b) 4 cm (c) $184\,\text{cm}^2$

T3 (a)

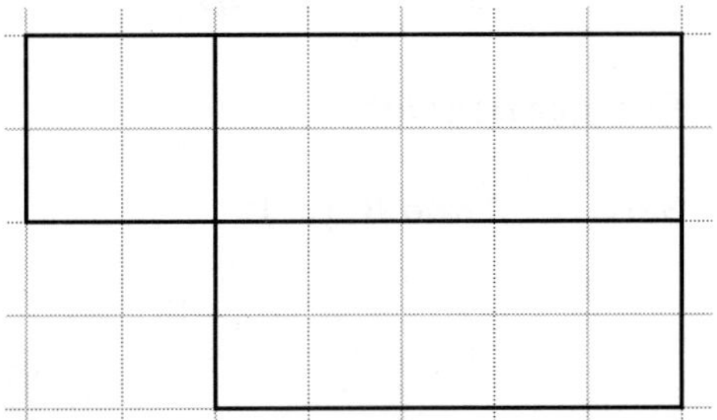

(b)

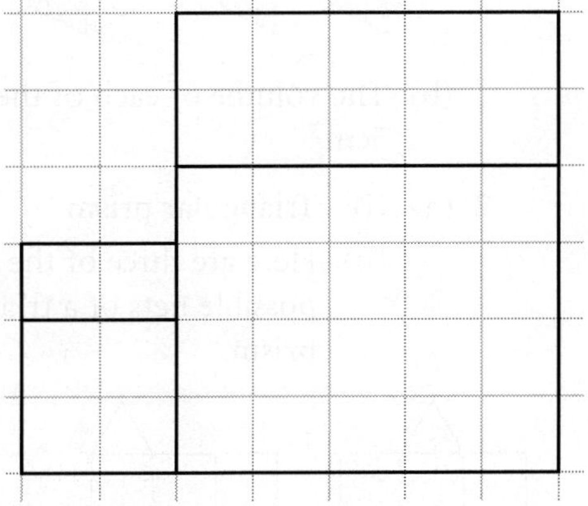

T4 A plane of symmetry shown on each shape.

Soma Cube solution

Practice booklet

Sections A and B (p 24)

1 (a)

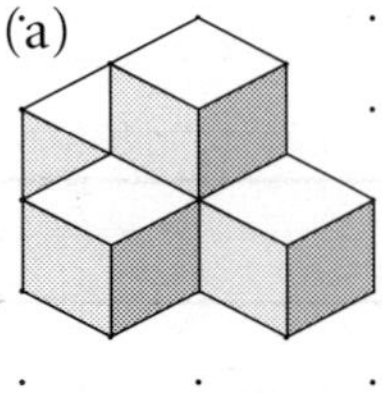

(b)

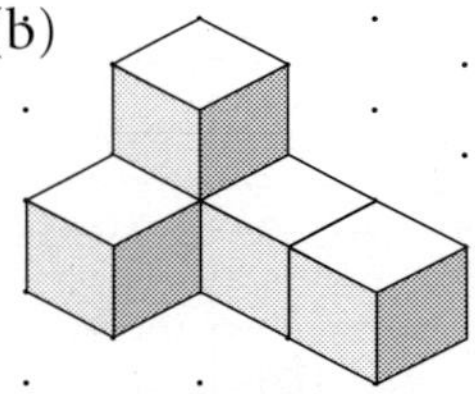

(b) The volume of each of the shapes is $5 cm^3$

2 (a) (i) Triangular prism

(ii) Here are three of the many possible nets of a triangular prism.

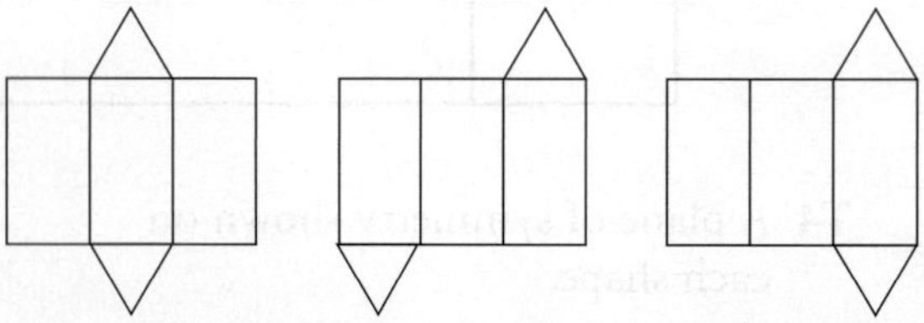

(b) (i) Hexagonal prism

(ii) Here are four of the many possible nets for a hexagonal prism.

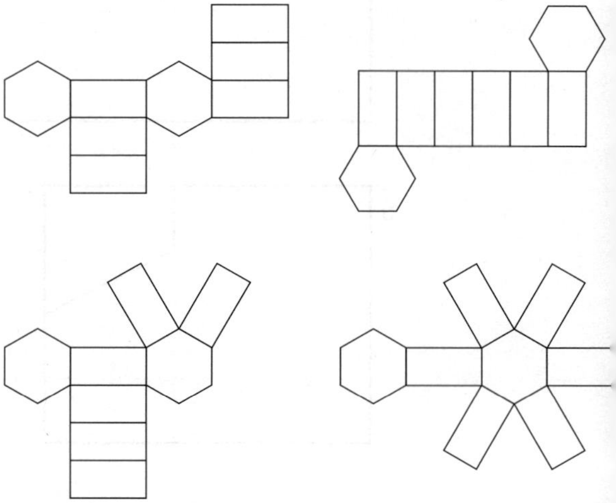

Section C (p 25)

1 (a) Plan

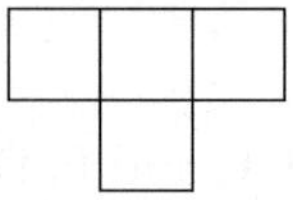

(b) Side

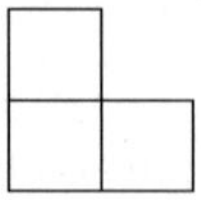

(c) Front

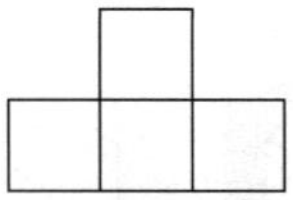

2 (a) plan

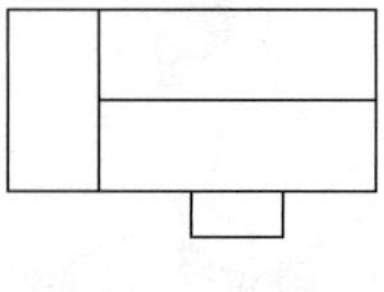

(b)

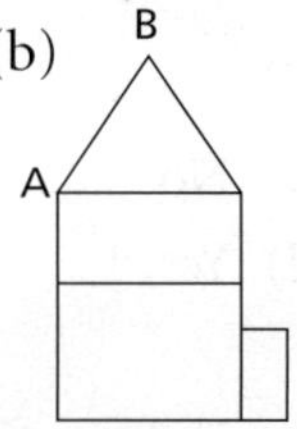

side

(c)

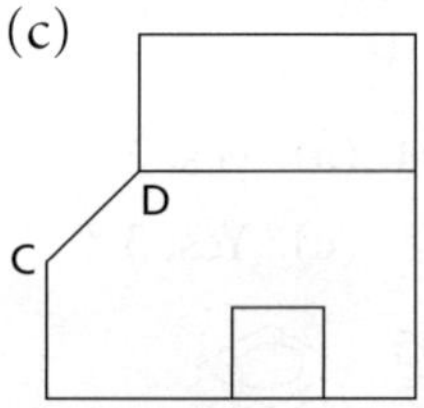

front

(d) Length of sloping edge AB = 3.6 m
Length of sloping edge CD = 2.8 m

Section D (p 26)

1 (a)

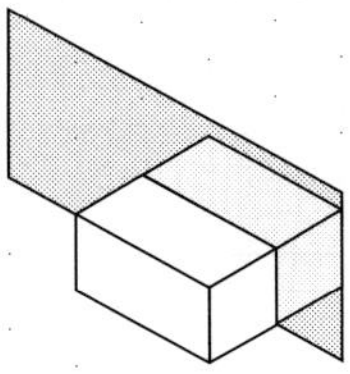

(b)

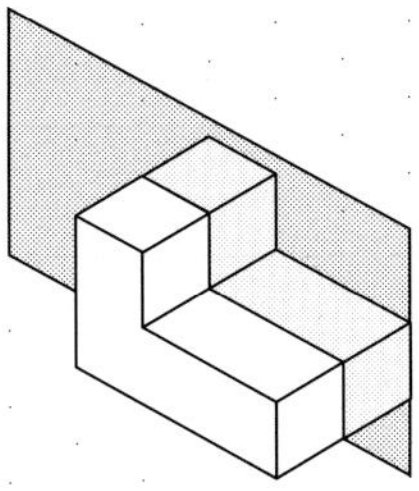

3 (a)

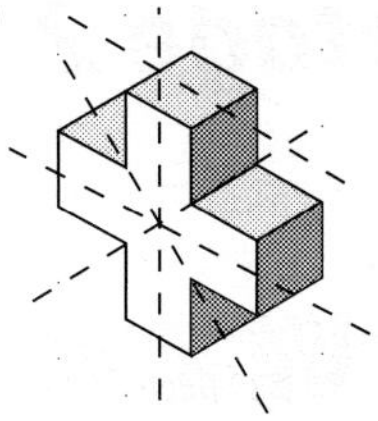

5 planes of symmetry

(b)

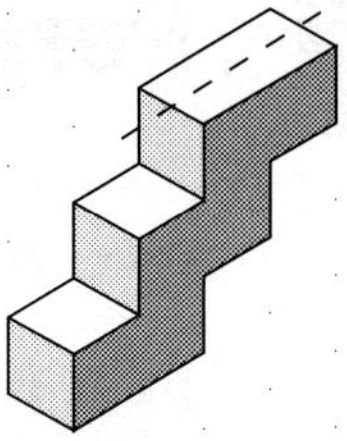

1 plane of symmetry

9 Percentage 2

Practice booklet pages 27 to 31

A ***Review*** (p 79)

A1 (a) £15 (b) 75%

A2 6

A3 (a) £60 (b) £6 (c) £15 (d) £10

A4 36

A5 (a) 0.25 (b) 0.72
(c) 0.6 or 0.60 (d) 0.07 (e) 0.02

A6 (a) 94% (b) 54% (c) 8%
(d) 80% (e) 1%

A7 62.4 grams

A8 (a) £9.80 (b) £35.60
(c) 69 (d) 201.6

A9 (a) 28% (b) 7300

A10 (a) £1.25 (b) £2.59
(c) £18.35 (d) £1.46

A11 (a) 0.265 (b) 0.638 (c) 0.502
(d) 0.013 (e) 0.058

A12 (a) 43.7% (b) 51.4% (c) 60.9%
(d) 5.2% (e) 2.9%

A13 (a) 15.7 kg (b) 49.7 m (c) 2.5

B ***Increasing and decreasing*** (p 80)

B1 (a) £6 (b) £54

B2 (a) 1 kg (b) 5 kg

B3 2300

B4 64 kg

B5 92p

B6 (a) 24 (b) 18p

C ***Increasing using multipliers*** (p 81)

C1 (a) £53.25 (b) £8.50 (c) £31.25
(d) £44.25 (e) £2.75

C2 1.12

C3 (a) £3.25 (b) £43.23 (c) £67.20
(d) £17.81 (e) £41.16

C4 £3.88

C5 189

C6 (a) Tim (b) 60 480

C7 1590

C8 £329.60

C9 £336.05

C10 £1445

C11 £314.40

C12 (a) £605.88 (b) £1010.40
(c) £15.38 (d) £3.84

D Decreasing using multipliers (p 82)

D1 (a) £14.28 (b) £4.76 (c) £76.50
(d) £11.04 (e) £0.17

D2 0.65

D3 (a) £16.25 (b) £19.24 (c) £32.50
(d) £32.50 (e) £25.99

D4 (a) £4.03 (b) £5.29 (c) £16.79
(d) £29.82 (e) £235.20

D5 0.91

D6 (a) £1128.40 (b) £618.80 (c) £8.65
(d) £0.58 (e) £13.38

D7 (a) 0.28 (b) 0.77 (c) 0.7
(d) 0.95 (e) 0.98

D8 (a) £208.80 (b) £111.60
(c) £2.30 (d) £395.60

D9 £5766

D10 A × 1.76 B × 0.76 C × 1.24
D × 0.24 E × 1.2 F × 0.8

D11 £119 900

D12 £3690

D13 30 kg

D14 7360

E Finding an increase as a percentage (p 84)

E1 8%

E2 (a) 14% (b) 65% (c) 70% (d) 7%

E3 30%

E4 85%

E5 21%

E6 32%

E7 (a) 92% (b) 9%

E8 England 29% Scotland 5%

*E9 23.9%

F Finding a decrease as a percentage (p 85)

F1 35%

F2 (a) 24% (b) 62% (c) 30% (d) 4%

F3 7%

F4 12%

F5 35%

F6 (a) 15% decrease (b) 43% increase
(c) 30% increase (d) 27% decrease
(e) 40% decrease (f) 9% increase
(g) 2% decrease (h) 7% increase

F7 (a) 21% decrease (b) 28% increase

F8

	% change
Cheese	16% increase
Beef and veal	43% decrease
Fish	12% decrease
Fresh green veg	40% decrease
Fresh fruit	21% increase
Instant coffee	86% increase

G Mixed questions (p 86)

G1 (a) £247.50 (b) £256 (c) £246.75

G2 £77.35

G3 7%

G4 (a) 166.4 cm (b) 14%

G5 49%

Test yourself (p 87)

T1 45

T2 £91

T3 2.17 m

T4 132

T5 3 253 000

T6 35%

T7 11%

Practice booklet

Sections A and B (p 27)

1 (a) 8 (b) 60 (c) 15%

2 (a) £10 (b) £6 (c) £2 (d) £32

3 (a) 0.75 (b) 0.48 (c) 0.8
(d) 0.01 (e) 0.08

4 (a) 87% (b) 62% (c) 40% (d) 3%

5 (a) £16 (b) £64

6 (a) 15 mm (b) 75 mm

7 £42

8 £166.60

9 0.089

10 (a) £16.06 (b) £2.51 (c) £1.34
(d) £112.39

11 81.4%

12 1105

Sections C and D (p 28)

1 1.05

2 (a) £36.80 (b) £907.20 (c) £259.48
(d) £247.25

3 £12 190

4 0.73

5 (a) £32.40 (b) £460.35
(c) £143.23 (d) £469.70

6 441

7 £18 900

8 3694

9 69 kg

10 34 840

11 £780

12 38 150

Sections E and F (p 29)

1 (a) 18% (b) 40% (c) 24.8 or 25%

2 7%

3 (a) 15% (b) 44% (c) 65%

4 5%

5 (a) 17% decrease (b) 9% decrease
(c) 19% decrease (d) 10% increase
(e) 3% decrease (f) 6% increase

6 7%

7 27%

8 28%

*__9__ (a) 21.7% decrease (b) 8.2% increase
(c) 57.3% increase

Section G (p 30)

1 (a) £52 (b) £19.25

2 (a) £411.25 (b) £190.35

3 (a) 2 200 000 (b) 2 400 000

4 (a) (i) 76% (ii) 97%
(b) (i) 72% (ii) 34%
(c) 9 300
(d) 12 800

10 Maps and plans

This unit looks at various issues related to using maps and scaled drawings. Ratio scales are used and bearings are revised. Four figure grid references are also included which are required by some examination boards.

Essential	Optional
Sheets P100 to P104 Angle measurers	Magnetic compasses
Practice booklet pages 32 to 34	

B *Grid references* (p 89)

Sheet P100

◊ Many students will be familiar with 6 figure grid references from Geography or from outdoor activities and may wish to use these.

◊ You may prefer to use a local map with your own questions for this section and subsequent sections.

C *Scales* (p 89)

This section looks at changing scales expressed in the form 1:n into the more familiar 1 cm represents ....

Sheets P100, P101, P102

◊ The answers given here assume that distances are measured to the nearest centimetre. You may wish to encourage a greater degree of accuracy in student's measuring but this should not distract from the essential purpose of using scales.

D *Bearing up* (p 91)

Sheets P100, P103 Angle measurers

◊ Students will have met bearings at KS3 and should be aware of the fact that at any point a line going North can be drawn vertically from that point.

D5 This uses bearings from two points to locate a position. A boat issuing a Mayday call close to land would give its position in this way from any prominent features. This question also raises the idea of back bearings, i.e. that to travel back along a course the bearing is 180° different from the original course. Although this can be done by measuring from the map this could be a topic for discussion.

Test yourself (p 93)

Sheets P103, P104

A *Working to plans* (p 88)

A1 8 cm; 16 m

A2 (a) 10 m (b) 5 m (c) 14 m

A3 140 m^2

A4 51 m

A5 Student's drawing of classroom.

B *Grid references* (p 89)

B1 (a) 85 69 (b) 86 68

B2 (a) Gravelhole Wood
(b) Quarry
(c) Hare Hill (d) Upton Hall

C *Scales* (p 89)

C1 (a) 500 m (b) 0.5 km (c) 1 km
(d) *On the map 2 cm represents 1 km.*

C2 (a) 3 cm (b) 1.5 km

C3 (a) 6 cm; 3 km (b) 2 cm; 1 km
(c) 7 cm; 3.5 km (d) 5 cm; 2.5 km

C4 (a) 25 000 cm (b) 250 m
(c) 4 cm

C5 (a) 1 km (b) 2 km (c) 3 km
(d) 2.5 km (e) 1.5 km (f) 1.75 km

C6 Fir Tree Farm and Withinlee Farm

C7 (a) 72 000 cm (b) 4 cm
(c) 18 000 cm (d) 1:18 000

C8 (a) 900 m (b) 2340 m
(c) 1620 m (d) 1980 m

C9 (a) 1 cm represents 2.5 km
(b) (i) 16 km (ii) 20 km
(iii) 10 km (iv) 13 or 14 km

C10 (a) 1:18 000 (Map C)
(b) 1:200 000 (Map D)
(c) Either 1:50 000 or 1:25 000 (Map A or B)

D Bearing up (p 91)

D1 (a) 070° (b) 148° (c) 231° (d) 322°

D2 (a) Bollington Cross
(b) Adder's Moss
(c) The Hollins
(d) Higherfence
(e) Mottram Cross
(f) Tidnock Wood

D3 (a) Thorneycroft Hall
(b) Crabtree Moss Farm

D4 (a) Journey marked: First stage goes to the Hollins in 92 72, second stage to Finlow Hill Farm in 85 76.
(b) Finlow Hill Farm
(c) 168° (d) 8 km

D5 (a) K marked in 00 06
(b) F marked in 96 03
(c) (i) 10° (ii) 190°
(iii) 345° (iv) 165°

D6 (a) 120° (b) 300°

D7 Point D marked in 96 08

Test yourself (p 93)

T1 (a) 6 cm (b) 3 km (c) 224°

T2 (a) (i) 252° (ii) 2.9 or 3 km
(b)

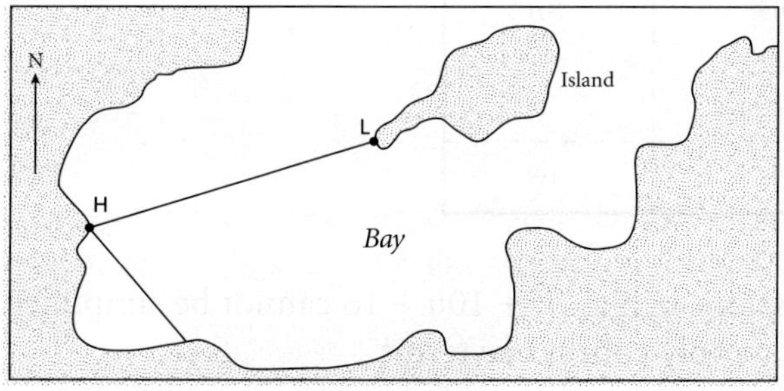

Practice booklet

Sections B and C (p 32)

1 (a) Henshaw Hall Farm
(b) Withinlee Farm
(c) Lodge Farm

2 (a) 1 km (b) 5 km (c) 1.5 km
(d) 12.5 km

3 (a) 400 m (b) 2.5 cm

4 (a) 150 km (b) 300 km (c) 135 km
(d) 408 km

5 (a) 2.5 km (b) 10 km (c) 25 km
(d) 37.5 km

6 (a) 8.6 cm (b) 4.3 km

7 (a) 2 cm (b) 12 cm (c) 17 cm

Section D (p 33)

1

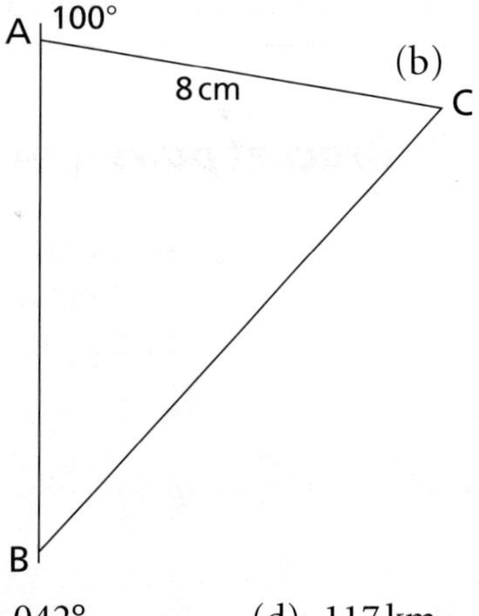

(c) 042° (d) 117 km

2 (a, b) An accurate copy of diagram with points O, A, B, C
(c) 9.2 km
(d) (i) 66.9° (ii) 130.1°

3 (a)

	A to B	B to C	C to D	D to E	E to A
Bearing	070°	118°	192°	292°	012°
Distance (m)	285	340	445	585	285

(b) 5820 m

11 Brackets and proof

TG

p 94	**A** *Bracket pairs*	Expanding brackets $(x + a)(x + b)$
p 95	**B** *Including subtraction*	Expanding brackets $(x \pm a)(x \pm b)$
p 96	**C** *Grid totals*	Finding general expressions for totals on grids
p 98	**D** *Opposite corners*	Making and proving general statements by manipulating expressions with brackets
p 99	**E** *Consecutive numbers*	Proving statements involving consecutive numbers
p 100	**F** *More brackets*	Expanding brackets $(ax \pm b)(cx \pm d)$ and $(ax \pm by)(cx \pm dy)$

Optional

Sheet P105

Practice booklet pages 35 to 38

A Bracket pairs (p 94)

TG

◊ In the student material, expanding brackets is linked to the 'table method' for long multiplication. Students may also find it helpful to link it to finding the area of a rectangle as shown below:

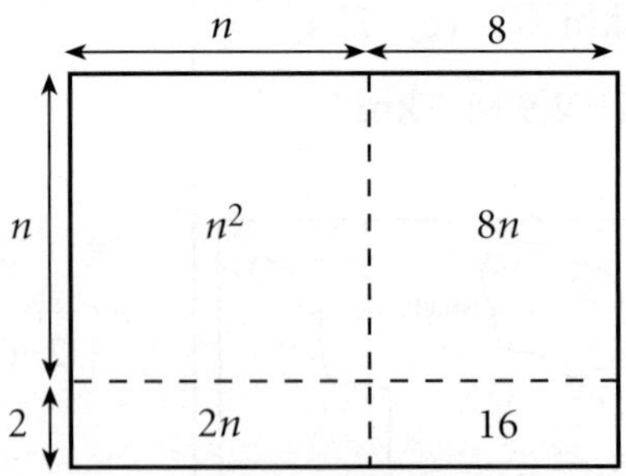

◊ Emphasise that an expression such as $n^2 + 10n + 16$ cannot be simplified further by adding. One can only collect *like* terms.

◊ Discuss the difference between the statement $(n + 2)(n + 8) = n^2 + 10n + 16$ and a statement such as $n^2 = 9$. Discuss the difference between an **identity** and an **equation**.

B *Including subtraction* (p 95)

TG

◊ Students can be asked to try multiplying out the brackets from the products A, B and C and their solutions and methods can be compared.

- One approach is to note the equivalence of $(x - 4)$ and $(x + {}^{-}4)$. Then use the table approach outlined on page 2 as follows:

$(x + {}^{-}4)(x + 5)$ ⇨

×	x	$^{-}4$
x	x^2	$^{-}4x$
5	$5x$	$^{-}20$

⇨

$$\begin{aligned}&(x - 4)(x + 5)\\&= (x + {}^{-}4)(x + 5)\\&= x^2 + 5x + {}^{-}4x + {}^{-}20\\&= x^2 + x - 20\end{aligned}$$

- Looking at areas of rectangles can be helpful when each bracket involves addition but can be confusing when one or both brackets involve subtraction. However, some may find it interesting to consider how **any** product of two brackets can be represented as an area.

C *Grid totals* (p 96)

Optional: Sheet P105 (blank 10-column grids)

C1 Some students may solve parts (e) and (f) by trial and improvement.

Point out that the numbers in part (e) can be found by solving the equation $3n + 11 = 614$ to give $n = 201$.

In (f), students can try to solve $3n + 11 = 154$ to find that n is a non-integer and hence the L-shape can't be found.

D *Opposite corners* (p 98)

Optional: Sheet P105

This section gives the opportunity for an extensive investigation.

◊ Discuss with students why trying squares in many different positions does not amount to a proof but an algebraic argument proves a statement true in **all** possible cases.

Many students will find this difficult to grasp. They may find it illuminating to look at the expression '$n^2 + n + 41$'. The expression is prime until $n = 40$ and students may decide it is **always** prime on the basis of trying many values less than 40. But $n = 41$ will prove them wrong.

D4 Some students may see that for an s by s square grid, the opposite corners number is $10(s - 1)^2$ although they may express this in a less formal way.

Only very few students will be able to attempt a proof at this level and it may be found easier after Section F.

For an s by s square with $x = s - 1$ we have a grid with corners n, $n + x$, $n + 10x$ and $n + 11x$ leading to a total of $10x^2$ i.e. $10(s - 1)^2$.

***D5** High attainers can think about the general rule that links the 'opposite corners number' with the size of the square **and** the number of columns.

For an s by s square on a grid with c columns the total is $c(s - 1)^2$.

A *Bracket pairs* (p 94)

A1 (a) $n^2 + 8n + 12$ (b) $n^2 + 13n + 12$
(c) $n^2 + 5n + 4$ (d) $n^2 + 6n + 8$
(e) $n^2 + 9n + 18$ (f) $n^2 + 11n + 18$

A2 $(x + 3)^2 = (x + 3)(x + 3)$
$= x^2 + 3x + 3x + 9$
$= x^2 + 6x + 9$

A3 (a) $x^2 + 8x + 16$ (b) $x^2 + 10x + 25$
(c) $x^2 + 2x + 1$

A4 Area is $(x + 3)(x + 5)$
$= x^2 + 5x + 3x + 15$
$= x^2 + 8x + 15$ as required

A5 (a) $(x + 3)(x + 4)$ $x^2 + 7x + 12$
(b) $(x + 1)(x + 8)$ $x^2 + 9x + 8$
(c) $(x + 2)^2$ $x^2 + 4x + 4$

A6 (a) $n + 4$ and $n + 3$
(b) $n + 5$ and $n + 1$
(c) $n + 6$ and $n + 1$
(d) $n + 2$ and $n + 3$

A7 (a) $n + 1$ and $n + 8$
(b) $n + 4$ and $n + 5$
(c) $n + 4$ and $n + 10$
(d) $n + 5$ and $n + 8$

B *Including subtraction* (p 95)

B1 (a) $n^2 + 3n - 10$ (b) $n^2 + 8n - 9$
(c) $n^2 - 4n - 12$ (d) $n^2 - 5n - 14$
(e) $n^2 - 7n + 12$ (f) $n^2 - 8n + 15$
(g) $n^2 - 8n + 7$ (h) $n^2 - n - 90$
(i) $n^2 - 16$ (j) $n^2 - 4$
(k) $n^2 - 25$ (l) $n^2 - 1$

B2 Area is $(x - 2)(x + 6)$
$= x^2 + 6x - 2x - 12$
$= x^2 + 4x - 12$ as required

B3 $(x - 4)^2 = (x - 4)(x - 4)$
$= x^2 - 4x - 4x + 16$
$= x^2 - 8x + 16$

B4 (a) $x^2 - 6x + 9$ (b) $x^2 - 2x + 1$
(c) $x^2 - 12x + 36$

B5 (a) $n + 3$ and $n - 1$
(b) $n + 1$ and $n - 2$
(c) $n - 2$ and $n - 3$
(d) $n - 1$ and $n - 3$
(e) $n - 2$ and $n + 2$
(f) $n - 3$ and $n + 3$

C *Grid totals* (p 96)

C1 (a) 47

(b) Student's different totals

(c) (i)

106	107
116	

(ii) 329

(d) (i)

n	$n+1$
$n+10$	

(ii) $3n + 11$

(e) 201, 202 and 211

(f) Student's explanation such as:
'$3n + 11 = 154$ leads to $n = 143 \div 3$ which is not a whole number.
So this L-shape is impossible.'

C2 (a) 105

(b) Student's different totals

(c) (i)

102	103	104
	113	
	123	

(ii) 545

(d) (i)

$n-1$	n	$n+1$
	$n+10$	
	$n+20$	

(ii) $5n + 30$

(e)

197	198	199
	208	
	218	

(f) Student's explanation such as:
'$5n + 30 = 231$ leads to $n = 201 \div 5$ which is not a whole number.
So this T-shape is impossible.'

C3 (a) (i)

$n-1$	n	$n+1$
	$n+6$	
	$n+12$	

(ii) $5n + 18$

(b) Student's investigation: the total for a T-shape on a grid with m columns is $5n + 3m$.

D *Opposite corners* (p 98)

D1 The 'opposite corners number' is always 40.

D2 (a)

n		$n+2$
$n+20$		$n+22$

(b) $(n + 2)(n + 20) = n^2 + 22n + 40$
$n(n + 22) = n^2 + 22n$
So the 'opposite corners number' is
$n^2 + 22n + 40 - (n^2 + 22n) = 40$.

(c) It proves that the 'opposite corners number' is 40 for **all** values of n.

D3 (a) Student's investigation

(b) The 'opposite corners number' is always 10.

(c) Any 2 by 2 grid is of the form:

n	$n+1$
$n+10$	$n+11$

$(n + 1)(n + 10) = n^2 + 11n + 10$
$n(n + 11) = n^2 + 11n$
So the 'opposite corners number' is
$n^2 + 11n + 10 - (n^2 + 11n) = 10$
proving that the 'opposite corners number' is 10 for **all** values of n.

D4 Student's investigation – the table below shows the opposite corners number for some different sizes of grid:

Size of square	'Opposite corners total'
2 by 2	10
3 by 3	40
4 by 4	90
5 by 5	160

***D5** Student's investigation

E *Consecutive numbers* (p 99)

E1 (a) The first number is $x - 1$ and the third number is $x + 1$.

(b) $x - 1 + x + x + 1$ simplifies to $3x$

(c) $3x = 3 \times x$ where x is a whole number so $3x$ **is** divisible by 3 ($3x \div 3 = x$).

E2 (a) (i) Student's three consecutive numbers

(ii) The product of the student's first and last numbers

(iii) The square of the student's middle number

(b) (i) Repeat of (a) with other sets of three consecutive numbers

(ii) The product of the first and third numbers is one less than the middle number squared.

(iii) One approach is to work on the same lines as **E1** and let the three numbers be $x - 1$, x and $x + 1$. The product of the first and third is $(x - 1)(x + 1)$ which expands and simplifies to $x^2 - 1$. The middle number squared is x^2 so the product of the first and third numbers will **always** be one less than the middle number squared.

An alternative approach is to let the three numbers be x, $x + 1$ and $x + 2$. The product of the first and third is $x(x + 2)$ which expands and simplifies to $x^2 + 2x$. The middle number squared is $(x + 1)^2$ which simplifies to $x^2 + 2x + 1$ so the product of the first and third numbers will **always** be one less than the middle number squared.

E3 One approach is to let the five numbers be $x - 2$, $x - 1$, x, $x + 1$ and $x + 2$. Then the sum is $5x$ and it can be easily seen that the sum is always five times the middle number.

An alternative approach is to let the five numbers be x, $x + 1$, $x + 2$, $x + 3$ and $x + 4$. Then the sum is $5x + 10$ and five times the middle number is $5(x + 2) = 5x + 10$.

E4 (a) $11 \times 12 - 10 \times 13$

(b) (i) $(n + 1)(\boldsymbol{n + 2}) - n(\boldsymbol{n + 3})$

(ii) $n^2 + 2n + n + 2 - (n^2 + 3n)$
$= n^2 + 3n + 2 - n^2 - 3n$
$= 2$
which proves that the value of each line of the pattern is 2.

***E5** (a) (i) 7 (ii) 17 (iii) 19

(b) Two consecutive square numbers can be written n^2 and $(n + 1)^2$. $(n + 1)^2$ expands to $n^2 + 2n + 1$ so the difference is $n^2 + 2n + 1 - n^2$ which is $2n + 1$. Since n is a whole number then $2n + 1$ is odd.

F *More brackets* (p 100)

F1 (a) $2x^2 + 7x + 3$ (b) $3x^2 + 7x + 2$
(c) $4x^2 + 11x + 6$ (d) $3x^2 + 13x - 10$
(e) $2x^2 - x - 3$ (f) $5x^2 + 18x - 8$
(g) $2x^2 - 15x + 25$ (h) $3x^2 - 5x + 2$
(i) $5x^2 - 12x + 4$

F2 (a) $6n^2 + 11n + 4$
(b) $9n^2 + 9n + 2$
(c) $15n^2 + 22n + 8$
(d) $25n^2 + 15n + 2$
(e) $8n^2 + 6n - 5$
(f) $6n^2 - 5n - 4$
(g) $10n^2 + 11n - 6$
(h) $4n^2 - 1$
(i) $6n^2 - 19n + 15$
(j) $10n^2 - 17n + 3$
(k) $14n^2 - 25n + 6$
(l) $24n^2 - 38n + 15$

F3 (a) $4b^2 + 4b + 1$
(b) $9k^2 + 12k + 4$
(c) $25h^2 - 20n + 4$

F4 The area of the whole rectangle is $(3x + 2)(x + 2)$ which expands to $3x^2 + 6x + 2x + 4$ which simplifies to $3x^2 + 8x + 4$.

Subtracting the pond area gives $3x^2 + 8x + 4 - 4 = 3x^2 + 8x$ as required.

F5 (a) $2x^2 + 5xy + 3y^2$
(b) $2x^2 + 5xy + 2y^2$
(c) $xy + 2x + 3y + 6$
(d) $3x^2 - 5xy - 2y^2$
(e) $2x^2 + 2xy - x - y$
(f) $2x^2 + 7xy - 15y^2$
(g) $2x^2 - 3xy + y^2$
(h) $5xy - x^2 - 6y^2$
(i) $16x - 4xy + 20 - 5y$
(j) $25x^2 - 4y^2$
(k) $4y^2 + 4xy + x^2$
(l) $4y^2 - 12xy + 9x^2$

Test yourself (p 101)

T1 (a) $x^2 + 3x + 2$ (b) $n^2 + 9n + 20$
(c) $a^2 + 2a - 15$ (d) $y^2 - 5y - 6$
(e) $k^2 - 6k + 8$ (f) $x^2 - 36$
(g) $n^2 + 12n + 36$ (h) $m^2 - 4m + 4$
(i) $p^2 - 10p + 25$

T2 (a) $2a^2 + 5a - 3$ (b) $3x^2 + 13x - 10$
(c) $2n^2 - 5n - 3$ (d) $3m^2 - 10m - 8$
(e) $2x^2 - 7x + 3$ (f) $6b^2 + 7b - 20$
(g) $6x^2 - x - 2$ (h) $10n^2 + 13n - 3$
(i) $9k^2 + 12k + 4$

T3 (a) 115
(b)

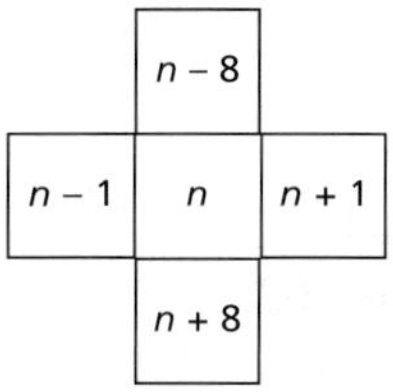

(c) $5n$
(d) As $C_n = 5n$, C_n is always a multiple of 5 and 134 is not a multiple of 5.

T4 (a) $2x^2 - 3xy - 2y^2$
(b) $6x^2 - 7xy - 5y^2$
(c) $9x^2 + 6xy + y^2$

***T5** Two consecutive numbers can be written n and $n + 1$ giving a sum of $2n + 1$.

$2n$ is even and adding 1 to an even number gives an odd number.

Practice booklet

Sections A and B (p 35)

1 (a) $p^2 + 5p + 6$ (b) $a^2 - 3a + 2$
(c) $a^2 + 3a - 10$ (d) $x^2 + 5x + 4$
(e) $y^2 - 4y - 12$ (f) $b^2 - 8b + 15$

2 (a) $a + 3$ and $a + 2$
(b) $a - 3$ and $a - 2$
(c) $a + 3$ and $a - 2$
(d) $a - 3$ and $a + 2$
(e) $a - 3$ and $a - 4$
(f) $a - 6$ and $a - 2$

3 (a) $p^2 + 6p + 8$ (b) $q^2 + 5q + 6$

4 (a) $s^2 + s - 2$ (b) $q^2 + q - 6$

Sections C, D and E (p 36)

1 (a) (i)

A

42	43
52	
62	63

B

77	78
87	
97	98

(ii) Totals 262 437

(b) (i)

n	$n+1$
$n+10$	
$n+20$	$n+21$

(ii) $5n + 52$
(iii) $n = 100$ so

100	101
110	
120	121

(iv) You would need to solve $5n + 52 = 100$ giving $n = 9.6$ which is not a whole number.

2 (a)

n	$n+1$
$n+8$	
$n+16$	$n+17$

(b) $5n + 42$

3 (a) $\boldsymbol{n - 1}, n, \boldsymbol{n + 1}, \boldsymbol{n + 2}$
(b) $4n + 2$
(c) $\frac{4n+2}{2} = 2n + 1$ which is a whole number so $4n + 2$ is divisible by 2.

4 (a) (i) (ii) (iii) Student's results
(b) (i) Student's results
(ii) The products differ by 2 each time.
(iii) If we call the numbers n, $n + 1$, $n + 2$ and $n + 3$ the products are $n(n + 3)$ and $(n + 1)(n + 2)$.
$n(n + 3) = n^2 + 3n$ and
$(n + 1)(n + 2) = n^2 + 3n + 2$
showing that the result for the two middle numbers is always 2 more than the result for the first and last numbers.

5 (a) $8 \times 9 - 4 \times 13$
$9 \times 10 - 5 \times 14$
(b) $(n + 4)(\boldsymbol{n + 5}) - n(\boldsymbol{n + 9})$
(c) $(n + 4)(n + 5) - n(n + 9)$
$= (n^2 + 9n + 20) - (n^2 + 9n)$
$= 20$
(d) This proves that the value of each line of the pattern is 20.

Section F (p 37)

1 (a) $2p^2 + 13p + 21$
(b) $3p^2 + 16p + 16$
(c) $4a^2 - a - 5$
(d) $3x^2 - 13x + 4$
(e) $9x^2 + 30x + 25$
(f) $4x^2 + 12x + 9$

2 (a) $15s^2 + 22s + 8$
(b) $8y^2 + 22y + 5$
(c) $14x^2 + 15x - 9$
(d) $10t^2 - 33t + 20$

3 (a) (i) $4x + 5$

(ii) $x + 4$
(iii) $4x^2 + 21x + 20$

(b) (i) Area of windows is $x \times 4x = 4x^2$ and so the area of the wall is $4x^2 + 21x + 20 - 4x^2 = 21x + 20$
(ii) $57.8\,m^2$

Review 2 (Student's book p 102)

Resource sheet P106 is needed for question 16

1 £12.74

2 $1\,512\,000 = 1.512 \times 10^6$
$201\,000\,000\,000 = 2.01 \times 10^{11}$

3 Area = length × width
$= (x + 7)(x + 2)$
$= x^2 + 2x + 7x + 14$
$= x^2 + 9x + 14$ as required

4 £7800

5 (a) 2 (b) 3 cm (c) $72\,cm^3$

(d) Full size sketches of these:

(i)

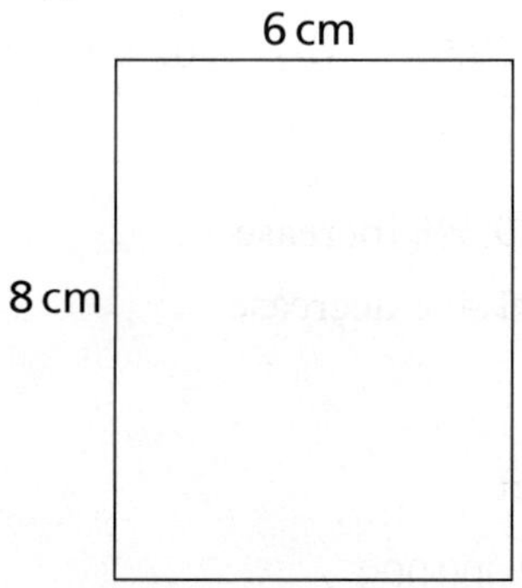

(ii)

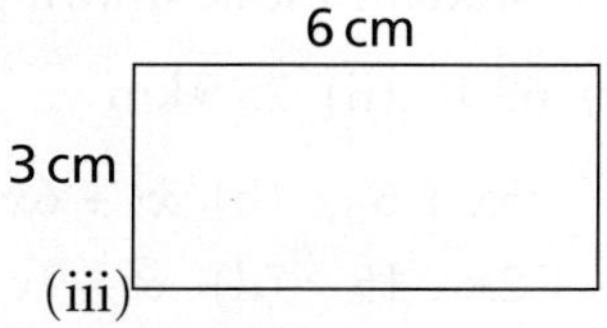

(iii)

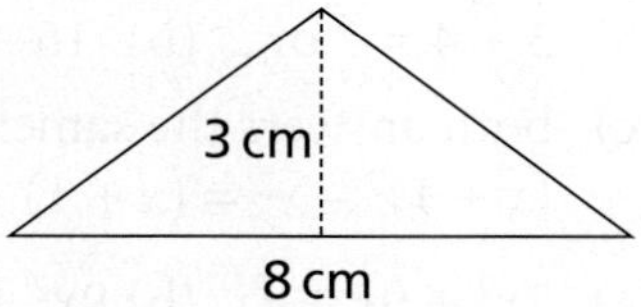

(e) Full size sketch of this:

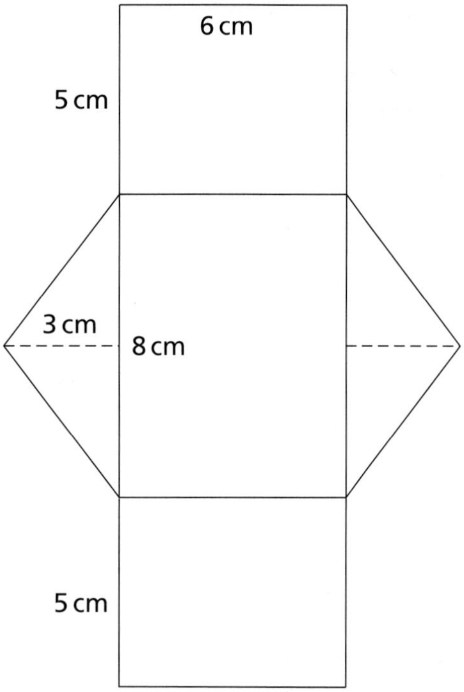

Surface area is $132\,cm^2$

6 £4648

7 0.00071 cm

8 £293.75

9 (a) $n^2 + 6n + 5$ (b) $x^2 + 2x - 8$
(c) $a^2 - 10a + 25$

10 3% or 3.4%

11 (a) Full size sketch of this:

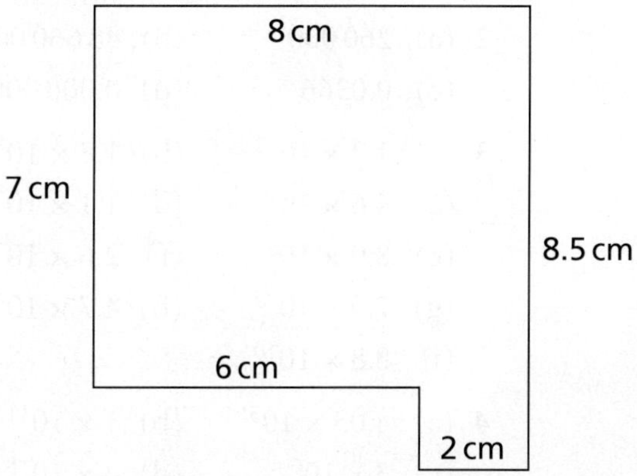

(b) 5.3 m (10.6 cm on sketch)

12 30%

13 1.1×10^{-22} grams

14 19%

15 26%

16 (a) 21.6 km

(b) (i) 115° (ii) 239°

(c) (i)

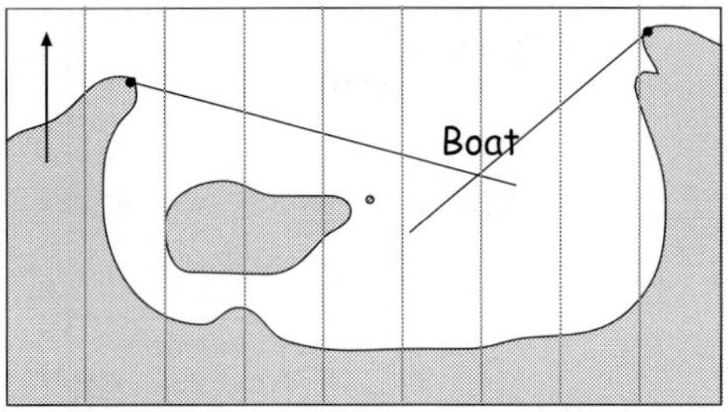

(ii) 4.6 km

17 (a) 3.2×10^{11} (b) 2×10^{4}

(c) 5×10^{-5}

18 One method is to label the first number as n giving n, $n + 1$ and $n + 2$ as three consecutive numbers.

The sum is $n + n + 1 + n + 2 = 3n + 3$ and three times the middle number is $3(n + 1) = 3n + 3$.

Mixed questions 2 (p 39)

1 (a) 4.5×10^{7} (b) 2.7×10^{7}

(c) 4.83×10^{-5} (d) 7.2×10^{-3}

2 (a) 260 000 (b) 48 660 000

(c) 0.0366 (d) 0.000 000 0486

3 (a) 1.2×10^{13} (b) 1.8×10^{-5}

(c) 4.6×10^{13} (d) 1.1×10^{-6}

(e) 8.0×10^{9} (f) 2.6×10^{-12}

(g) 7.9×10^{-7} (h) 4.7×10^{-11}

(i) 8.8×10^{10}

4 (a) 4.03×10^{5} (b) 1×10^{11}

(c) 3×10^{6} (d) 4×10^{-8}

(e) 1.6×10^{11} (f) 9×10^{8}

5 (a)

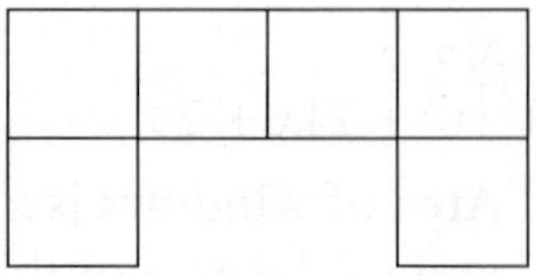

(b)

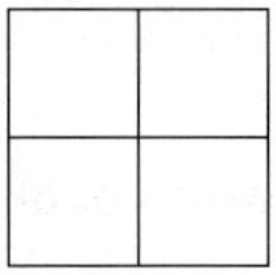

(c)

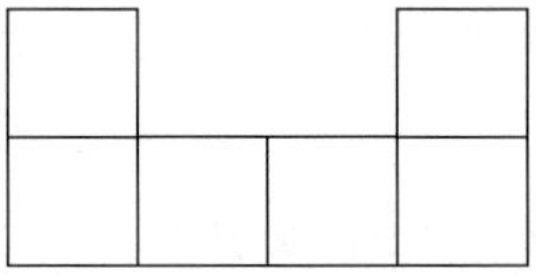

6

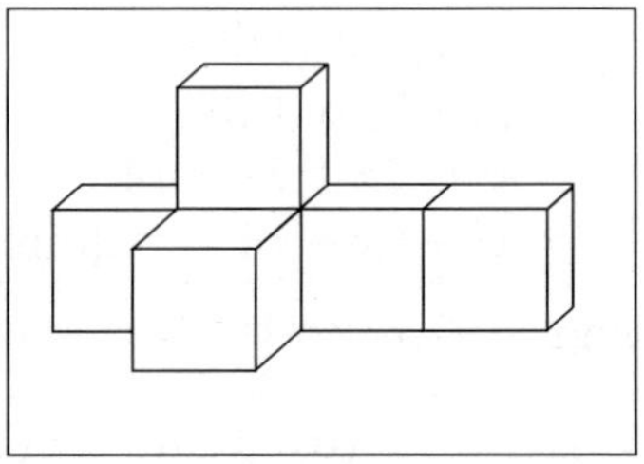

7 (a) £8.88 (b) £5.40

(c) £2.36 (d) £1.53

8 (a) £29.97 (b) £562.60 (c) £79.13

9 989

10 (a) 49.3% increase

(b) 21.4% decrease

11 8.3%

12 7.5 km

13 1 to 2 000 000

14 (a) The student's scale drawing

(b) (i) 69.1° (ii) 23.9 km

15 (a) $x^2 + 5x + 6$ (b) $x^2 + 6x + 9$

(c) $x^2 + 2x - 15$ (d) $x^2 - 9x + 20$

16 (a) $3 + 4 = 7$ or... (b) $16 - 9 = 7$

(c) both answers the same:
$(x + 1)^2 - x^2 = (x + 1) + x$

17 (a) $2x^2 + 9x + 4$ (b) $9x^2 + 12x + 4$

(c) $12x^2 - x - 1$ (d) $6x^2 - 13x + 6$

12 Ratio

The emphasis here is on the concept of ratio and its connections with fractions and percentages. Almost all the work can be done without using a calculator and thus provides plenty of practice in the use of basic number knowledge and skills.

TG

p 104 **A** *Ratio and recipes*	Revision of equivalent ratios, simplest form
p 106 **B** *Mixtures*	
p 108 **C** *Dividing in a given ratio*	
p 109 **D** *Ratios, fractions and percentages*	
p 110 **E** *k : 1 and 1 : k*	

Practice booklet pages 41 to 44

A *Ratio and recipes* (p 104)

◊ The opening activity should give you an opportunity to see how much students can remember about earlier work on ratio.

D *Ratios, fractions and percentages* (p 106)

◊ Given a ratio, there are a number of different comparisons that can be made using fractions. For example, if the ratio of boys to girls in a class is 3 : 2, we can link this ratio to several different fractions: the boys form $\frac{3}{5}$ of the class, the number of girls is $\frac{2}{3}$ of the number of boys, and so on. To avoid confusion it is useful to draw a simple diagram showing the information as given, from which other information can easily be read off. Examples are given in the text.

A *Ratio and recipes* (p 104)

A1 1 : 4, 20 : 80
6 : 10, 3 : 5
150 : 200, 3 : 4
25 : 75, 2 : 6

A2 (a) 5 : 2 (b) 2 : 5

A3 (a) 1 : 4 (b) 1 : 3 (c) 3 : 4 (d) 4 : 5
(e) 2 : 1 (f) 9 : 10 (g) 3 : 2 (h) 8 : 5
(i) 3 : 2 (j) 7 : 5 (k) 3 : 1 (l) 6 : 7
(m) 3 : 8 (n) 5 : 2 (o) 3 : 7

A4 (a) 2 : 5 (b) 2 : 5 (c) 2 : 5

A5 (a) Because one measurement is in cm and the other in m
(b) 1 : 50

A6 (a) 1 : 5 (b) 3 : 10 (c) 1 : 5
(d) 4 : 15 (e) 12 : 1 (f) 10 : 1
(g) 20 : 1 (h) 12 : 1

A7 3 : 1 : 2

A8 20 : 4 : 5

A9 20 : 5 : 2

A10 30 : 5 : 2

A11 (a) 1 : 4 : 6 (b) 2 : 3 : 5
(c) 2 : 3 : 5 (d) 4 : 3 : 2

A12 (a) 6 : 3 : 2 (b) A, C

B *Mixtures* (p 106)

B1

Red (1 part)	White (3 parts)
2 litres	**6 litres**
3 litres	9 litres
0.5 litre	**1.5 litres**
4 litres	12 litres

B2

Red (2 parts)	White (3 parts)
6 litres	**9 litres**
10 litres	15 litres
8 litres	**12 litres**
1 litre	1.5 litres

B3 (a) 2.5 litres (b) 100 ml

B4 (a) 40 (b) 25

B5 6

B6 64

B7 (a) 25 kg (b) 12 kg

B8

Ratio	Quantities
1 : 5	15 kg : **75** kg
2 : 3	50 g : **75** g
5 : 4	20 m : **16** m
3 : 8	**120** g : 320 g

B9 (a) £3.30 (b) £3.50

B10 (a) 80 (b) 15

B11 16

B12 7.5 litres

B13 12.5 litres

B14 (a) 48 ml mango juice, 72 ml fizzy water
(b) 250 ml pineapple, 100 ml mango

B15 90 cows, 36 pigs

B16 168

C *Dividing in a given ratio* (p 108)

C1 Colin £50, Dilip £150

C2 Emma £60, Farnaz £100

C3 (a) 600 ml or 0.6 litre
(b) 400 ml or 0.4 litre

C4 (a) 45 g (b) 5 g

C5 (a) £3, £9 (b) £10, £15
(c) 18 kg, 42 kg (d) £3.50, £1

C6 (a) 2 : 1 (b) 320

C7 1.4 kg

C8 Paul £4, Steph, £8, Ronnie £12

C9 Blue 0.25 litres, yellow 0.75 litres, white 1 litres

C10 10 chocolate, 10 cream, 15 plain

C11 140 g copper, 140 g nickel, 70 g zinc

D *Ratios, fractions and percentages* (p 109)

D1 $\frac{1}{5}$

D2 (a) $\frac{1}{8}$ (b) $\frac{7}{8}$

D3 (a) $\frac{3}{5}$ (b) 60%

D4 1 : 1

D5 1 : 2

D6 1: 3

D7 1 : 5

D8 3 : 2

D9 A, H B, D C, E F, G

D10 (a) $\frac{3}{5}$ (b) 60%

D11 A, E B, C The student's statement matching D, e.g. 'The ratio of boys to girls is 3 : 1'.

D12 4 : 1

E ***k* : 1 and 1 : *k*** (p 110)

E1 1.5 litres

E2 (a) 3.5 : 1 (b) 2.5 : 1 (c) 1.25 : 1
(d) 3.3 : 1 (e) 0.4 : 1

E3 1 : 5.5

E4 (a) 1 : 2.5 (b) 1 : 3.5 (c) 1 : 4.5
(d) 1 : 0.7 (e) 1 : 0.75

E5 1 : 19

E6 3.5 : 1

E7 (a) 11.5 : 1 (b) 575 g

Test yourself (p 111)

T1 20

T2 (a) 720 City, 480 United
(b) $\frac{1}{8}$

T3 (a) Sharon £234, Liam £126
(b) 65%

T4 (a) £735 (b) £196

Practice booklet

Section A (p 41)

1 (a) 7 : 3 (b) 3 : 7

2 1 : 5 = 20 : 100
75 : 25 = 3 : 1
5 : 4 = 25 : 20

3 (a) 1 : 4 (b) 1 : 5 (c) 3 : 1 (d) 3 : 4
(e) 7 : 6 (f) 6 : 7 (g) 4 : 9 (h) 20 : 7
(i) 8 : 5 (j) 5 : 3

4 (a) 1 : 3 (b) 4 : 1 (c) 5 : 8

5 1 : 4

6 (a) 3 : 5 (b) 1 : 4 (c) 1 : 4
(d) 5 : 1 (e) 1 : 5 (f) 20 :1
(g) 2 : 15 (h) 25 : 2

7 9 : 4 : 3

8 1 : 3 : 6, 5 : 15 : 30
25 : 75 : 100, 5 : 15 : 20
10 : 20 : 40, 3 : 6 : 12

9 6 : 3 : 2

Section B (p 42)

1

Blue (1 part)	White (4 parts)
2 litres	**8 litres**
3 litres	12 litres
0.5 litres	**2 litres**
2.5 litres	10 litres

2

Yellow (4 parts)	Blue (3 parts)
8 litres	**6 litres**
12 litres	9 litres
2 litres	**1.5 litres**
10 litres	7.5 litres

3 25 girls

4 180 red blocks

5

Ratio	Quantities
1 : 2	**200 m : 400 m**
2 : 5	10 kg : **25 kg**
4 : 7	16 cm : **28 cm**
10 : 9	**50 g** : 45 g

6 (a) 20 lobelia (b) 27 marigolds

7 15 litres peat, 10 litres sand

8 15 kg sand, 30 kg ballast

9 3.05 m

Section C (p 43)

1 Tracy £60, May £90

2 (a) 250 g (b) 500 g

3 Kate 140 g, David 160 g

4 (a) 3 : 5
(b) James £15, Naomi £25

5 (a) £4, £20 (b) 20 m, 12 m
(c) 10 kg, 35 kg (d) £7.20, £4.80

6 (a) 36 coloured pages
(b) 108 black and white pages

7 Mark £40, Kerry £60, John £20

8 150 red, 90 blue, 60 yellow

9 120 adults

Section D (p 44)

1 (a) $\frac{1}{3}$ (b) $\frac{2}{3}$

2 1 : 4

3 (a) $\frac{1}{4}$ (b) 25 %

4 2 : 1

5 A and E, B and H, C and F, D and G

Section E (p 44)

1 0.75 litres

2 (a) 2.5 : 1 (b) 1.8 : 1 (c) 1.3 : 1
(d) 1.875 : 1 (e) 0.6 : 1

3 (a) 1 : 3.5 (b) 1 : 2.4 (c) 1 : 2.5
(d) 1 : 0.9 (e) 1 : 0.8

4 1 : 1.25

5 1.8 : 1

13 The tangent function

Practice booklet pages 45 to 47

A *Finding an 'opposite' side* (p 112)

TG

◊ The initial activity is designed to lead to awareness that the vertical length can be obtained by multiplying the base length by a constant amount, in this case about 0.7 (or the equivalent fact that vertical divided by horizontal gives 0.7).

An alternative approach is to get the class to draw a 35° right-angle triangle, but with each student given a different base length. Students are then asked to measure the vertical length and the results are put up on the board until the 'rule' emerges.

B *Finding an adjacent side* (p 115)

TG

◊ Finding the adjacent side of the triangle at the top of the page can be presented as a challenge for discussion in small groups. Two valid approaches may emerge: one is dividing the opposite side by 0.7 to get the adjacent side (the inverse of multiplying the adjacent side by 0.7 to get the opposite side); the other is to write the equation $\frac{8.4}{x} = 0.7$ and solve it.

D *Mixed questions and problems* (p 118)

◊ Stress the importance of drawing a sketch, drawing additional lines where necessary and marking quantities on the sketch as they are worked out.

A Finding an 'opposite' angle (p 112)

A1 (a) 7.7 cm (b) 9.1 cm (c) 21 cm
(d) 10.5 cm (e) 15.4 cm (f) 28 cm

A2 The student's drawing

A3 (a) 16.7 cm (b) 20.3 cm
(c) 21.5 cm (d) 29.8 cm

A4 adjacent side → × 1.19 → = opposite side

or $\frac{\text{opposite side}}{\text{adjacent side}} = 1.19$

A5

Angle	Adjacent side	× ?	Opposite side
10°	10 cm	0.16	1.6 cm
20°	10 cm	0.36	3.6 cm
30°	10 cm	0.56	5.8 cm
40°	10 cm	0.82	8.2 cm
50°	10 cm	0.12	11.7 cm
60°	10 cm	1.74	17.4 cm
70°	10 cm	2.75	27.5 cm

A6 (a) 5.9 cm (b) 4.6 cm
(c) 24.3 cm (c) 4.0 cm

A7 tan 10° = 0.176 tan 20° = 0.364
tan 30° = 0.577 tan 40° = 0.839
tan 50° = 1.192 tan 60° = 1.732
tan 70° = 2.747

A8 (a) 2.145 (b) 17.2 cm
(c) 9.01 cm (d) 471.7 cm

A9 (a) The adjacent and opposite sides are equal in length.
(b) 45°
(c) opposite divided by adjacent equals one, so tan 45° = 1

A10 (a) (i) The student's estimate
(ii) 1.5 cm
(b) (i) The student's estimate
(ii) 3.2 cm
(c) (i) The student's estimate
(ii) 5.8 cm
(d) (i) The student's estimate
(ii) 3.7 cm
(e) (i) The student's estimate
(ii) 3.0 cm
(f) (i) The student's estimate
(ii) 2.3 cm

B Finding an adjacent side (p 115)

B1 (a) 50 cm (b) 30 cm
(c) 17 cm (d) 14 cm

B2 (a) 20 cm (b) 16 cm
(c) 50 cm (d) 52 cm

B3 (a) (i) The student's estimate
(ii) 11.5 cm
(b) (i) The student's estimate
(ii) 8.8 cm
(c) (i) The student's estimate
(ii) 7.1 cm
(d) (i) The student's estimate
(ii) 9.1 cm
(e) (i) The student's estimate
(ii) 6.0 cm
(f) (i) The student's estimate
(ii) 4.4 cm

B4 (a) 7.6 cm (b) 20.9 cm
(c) 9.8 cm (d) 11.5 cm

C Finding an angle (p 116)

C1 (a) 0.42 (b) 38° (c) 1.28 (d) 76°
(e) 16° (f) 0.11 (g) 3° (h) 14.3
(i) 44° (j) 88°

C2 (a) 1.6 (b) 58°

C3 Tangents get larger very quickly as a approaches 90°.

C4 (a) $\tan a = 0.8$ $a = 39°$
(b) $\tan a = 0.4$ $a = 22°$
(c) $\tan a = 2.6$ $a = 69°$
(d) $\tan a = 0.9$ $a = 42°$

C5 (a) $\tan x = 0.45$ $x = 24°$
(b) $\tan x = 0.55$ $x = 29°$
(c) $\tan x = 4.7$ $x = 78°$

C6 (a) $\tan a = 0.8$ $a = 38.66°$
(b) $\tan a = 0.4$ $a = 21.80°$
(c) $\tan a = 2.6$ $a = 68.96°$
(d) $\tan x = 0.45$ $x = 24.23°$
(e) $\tan x = 0.55$ $x = 28.81°$
(f) $\tan x = 4.7$ $x = 77.99°$

C7 $a = 26.57°$ $b = 18.43°$ $c = 68.20°$
$d = 14.04°$

C8 $a = 36.87°$ $b = 51.34°$ $c = 29.74°$
$d = 50.19°$ $e = 23.96°$ $f = 49.40°$
$g = 68.96°$ $h = 59.04°$ $i = 40.60°$

C9 $a = 55.27°$ $b = 50.50°$ $c = 56.05°$
$d = 37.06°$ $e = 64.54°$ $f = 33.50°$
$g = 62.51°$

D *Mixed questions and problems* (p 118)

D1 (a) 19.1 cm (b) 10.9 cm (c) 35.5°
(d) 15.0 cm (e) 39.6° (f) 17.7 cm
(g) 20.3 cm

D2 30°

D3 16°

D4 (a) 51.2° (b) 231.2°

D5 2.6 m

D6 23 m

D7 (a) 320° (b) 140°

D8 71 m

D9 6.0 cm

D10 75.6°

D11 $x = 66.5°$ $y = 113.5°$

D12 115.4° and 64.6°

D13 (a) 8.66 cm (b) 43.30 cm^2

Test yourself (p 121)

T1 (a) AC (b) EF (c) GI (d) KM

T2 (a) BC (b) DF (c) HI (d) KL

T3 (a) 10.6 cm (b) 64.5°
(c) 10.7 cm (d) 34.3°

T4 17.3 cm

Practice booklet

Section A (p 45)

1 (a) 5.95 cm (b) 7.5 cm (c) 3.58 cm
(d) 38.36 cm (e) 16.78 cm (f) 9.48 cm
(g) 44.98 cm (h) 16.80 cm (i) 36.49 cm
(j) 6.62 cm

Section B (p 45)

1 (a) 11.52 cm (b) 14.38 cm (c) 12.28 cm
(d) 16.58 cm (e) 21.33 cm (f) 13.36 cm
(g) 19.23 cm (h) 12.87 cm (i) 26.99 cm
(j) 35.97 cm

2 (a) 29.4 cm (b) 13.44 cm (c) 81.45 cm
(d) 26.29 cm (e) 11.95 cm (f) 27.72 cm
(g) 26.11 cm (h) 17.30 cm (i) 8.81 cm
(j) 62.41 cm

Section C (p 46)

$a = 35.3°$ $b = 37.8°$ $c = 64.1°$
$d = 22.0°$ $e = 34.3°$ $f = 19.0°$
$g = 42.4°$ $h = 38.4°$ $i = 49.1°$
$j = 52°$ $k = 28.5°$ $l = 55.3°$
$m = 14.7°$ $n = 54.5°$ $o = 63.4°$

Section D (p 47)

1 1.59 m

2 60.26°

3 25.02°

4 114°

5 44.1 cm^2

6 217 cm or 2.17 m

14 Cumulative frequency

Essential

Watches (for timing seconds), graph paper

Practice booklet pages 48 to 51

A *Hold your breath!* (p 122)

TG

◊ Ask everyone, working in pairs, to find the length of time for which they can hold their breath.

On the board or OHP, record the results on a dot plot:

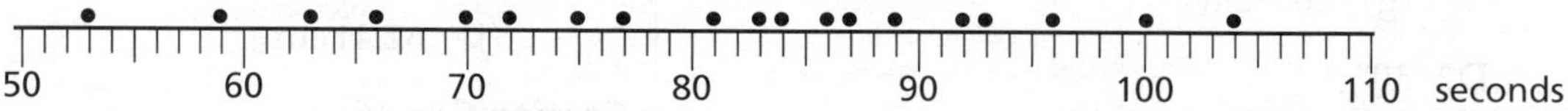

Decide on suitable class intervals (e.g. 50–60, 60–70, …) and record the number of students who held their breath for up to (and including), say, 60 seconds, 70 seconds, etc.

Time (t seconds)	Cumulative frequency
$t \leq 60$	2
$t \leq 70$	5

Then ask questions about this data, such as 'How many (or what percentage) held their breath for more than 90 seconds?'

C Cumulative frequency graphs (p 124)

◊ It is also permissible to join the points by straight line segments. This is equivalent to assuming that the distribution in each interval is uniform.

D Median, quartiles and inter-quartile range (p 125)

TG

◊ The graph at the foot of page 125 gives you an opportunity to ask students to think about how to find the median.

◊ Many students find it difficult to distinguish between what the range tells you and what the inter-quartile range tells you. In a population there may be one or two values at the extremes and these will define the range, even though the majority of the population may lie within a much narrower range either side of the median. The inter-quartile range tells you how spread out the 'middle half' of the population is, so it is generally a better basis for comparing spread than is the range.

E Box-and-whisker plots (p 128)

◊ Comparison of spread is well-illustrated by box-and-whisker plots.

For example, although population A below has a greater range than B, its inter-quartile range is smaller. So 'on the whole' B has a wider spread than A, as measured by the inter-quartile range.

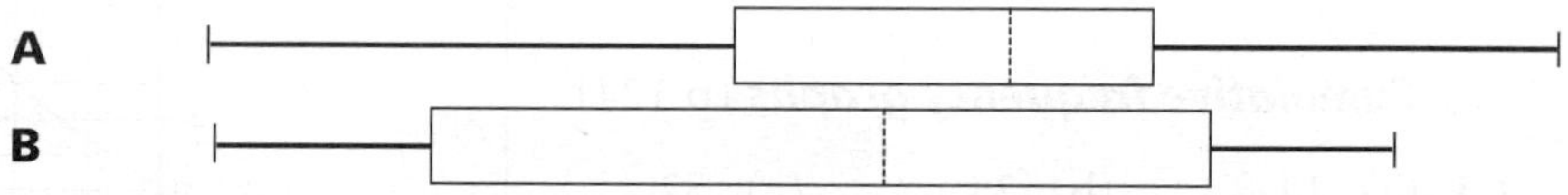

Practical work

The technique of drawing a cumulative frequency graph to find median, quartiles, etc, needs a large data set. The activity suggested below could be carried out in several classes and the data aggregated.

A student sits on a desk with their hands on the desk and one leg held out horizontally. They are timed for how long they can hold their leg in this position. They do it under two different conditions:

(1) They cannot see other people doing it while they are being timed.

(2) They can see others doing it.

Before starting, ask students to think about what sort of differences there might be between the two sets of data and how practically to carry out the experiment.

𝔹 ***Cumulative frequency tables*** (p 122)

B1 (a) 51, 60 (b) Total number of babies

B2 (a) 31 (b) 45 (c) 50 (d) 52

B3 (a) 21 (b) 7

B4

Height (h cm)	Cumulative frequency
$h \le 65$	2
$h \le 70$	11
$h \le 75$	23
$h \le 80$	29
$h \le 85$	33

B5 (a) 9 (b) 23 (c) 21 (d) 22

B6 (a)

Age (a years)	Cumulative frequency
$a \le 20$	4
$a \le 30$	9
$a \le 40$	19
$a \le 50$	26
$a \le 60$	28

(b)

Height (h cm)	Cumulative frequency
$h \le 150$	3
$h \le 160$	10
$h \le 170$	24
$h \le 180$	44
$h \le 190$	56

ℂ ***Cumulative frequency graphs*** (p 124)

C1 (a) 12 (b) 28 (c) 52
(d) 23 (e) 41 (f) 35

C2 about 24 people

C3 about 22 people

C4 40 seconds

C5 (a)

Waist (w cm)	Cumulative frequency
$w \le 65$	12
$w \le 70$	32
$w \le 75$	60
$w \le 80$	72
$w \le 85$	80

(b)

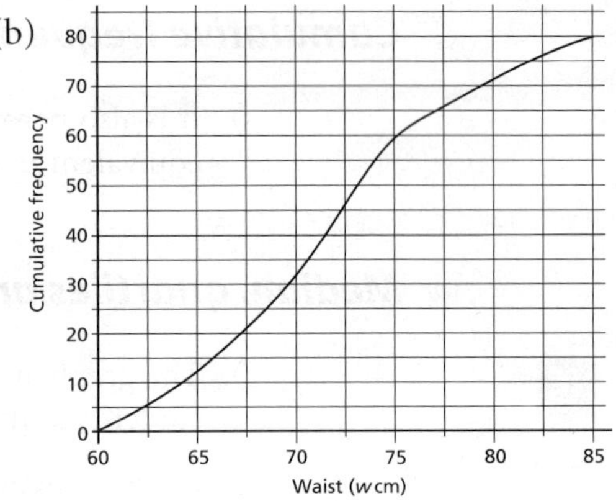

(c) 45 boys

C6 (a) 250

(b)

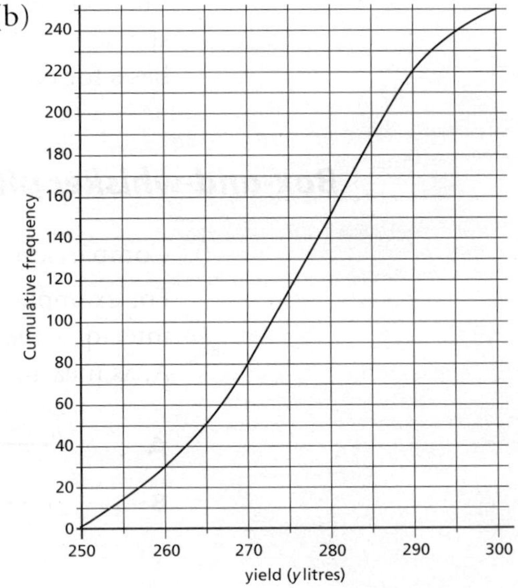

(c) (i) 188 cows (ii) 199 cows

(d) about 190

C7 (a)

Number of marks, m	Cumulative frequency
$m \le 10$	4
$m \le 20$	11
$m \le 30$	23
$m \le 40$	46
$m \le 50$	84
$m \le 60$	130
$m \le 70$	150
$m \le 80$	166
$m \le 90$	177
$m \le 100$	180

(b) 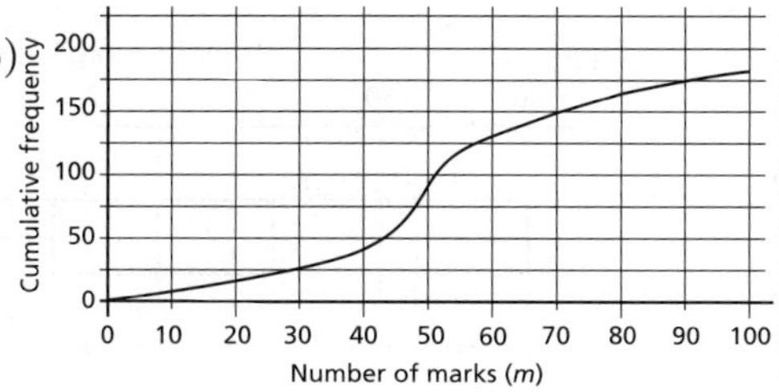

(c) 25 people (d) 60 people

D *Median, quartiles and inter-quartile range* (p 125)

D1 (a) 169 cm (b) 163 cm
(c) 177.5 cm (d) 14.5 cm

D2 (a) 25% have an armspan below 163 cm
(b) 50% have an armspan below 169 cm
(c) 75% have an armspan below 177.5 cm
(d) 25% have an armspan above 177.5 cm

D3 Median 44 mph, lower quartile 38 mph, upper quartile 48.5 mph

D4 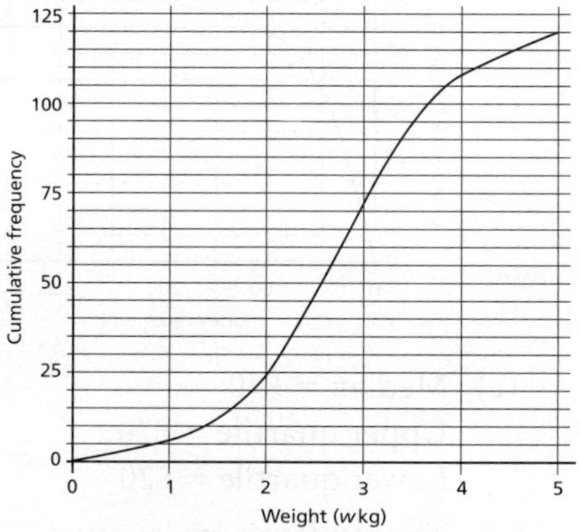

Median = 2.75 kg
Lower quartile = 2.2 kg
Upper quartile = 3.4 kg
Inter quartile range = 1.2 kg

D5 (a)

Weight (w g)	Cumulative percentage
$w \leq 60$	8%
$w \leq 70$	21%
$w \leq 80$	45%
$w \leq 90$	72%
$w \leq 100$	88%
$w \leq 110$	97%
$w \leq 120$	100%

(b) 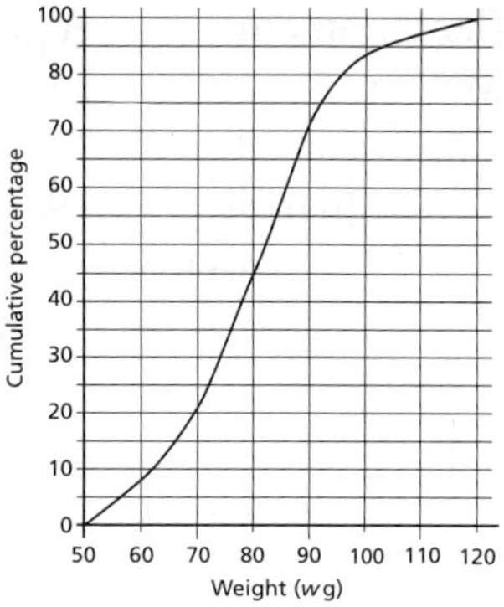

(c) Median = 82 g
Lower quartile = 71 g
Upper quartile = 92 g
Inter quartile range = 21 g

E *Box-and-whisker plots* (p 128)

E1 The length of the box

E2 (a) 38 years (b) 32 years
(c) 25% (d) 50%
(e) 42 years in town A and 38 years in town B
(f) Town B – median is higher
(g) There is very little difference. Town A has a slightly higher inter-quartile range but its range is lower.

E3 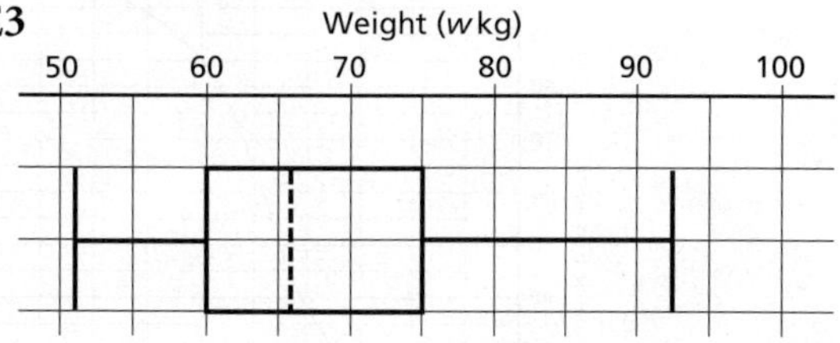

E4 The girls' handspans are more varied, and the boys handspans are generally larger.

E5 (a) E, because it has the greatest median.
(b) E, because it has the greatest inter-quartile range, or A, which has the greatest range.

E6 (a)

(length in cm)	A	B
Median	23	31
Upper quartile	27	34
Lower quartile	19	26

(b)

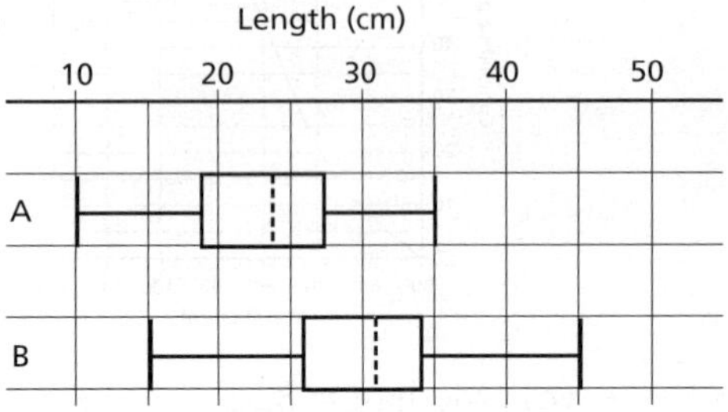

(c) 8 cm for both groups

(d) Although both groups have the same inter-quartile range, the worms in group B are on the whole larger and have a greater range of lengths.

Test yourself (p 130)

T1 (a)

Weight (w kg)	Cumulative frequency
$w \le 50$	15
$w \le 60$	37
$w \le 70$	85
$w \le 80$	110
$w \le 90$	120

(b)

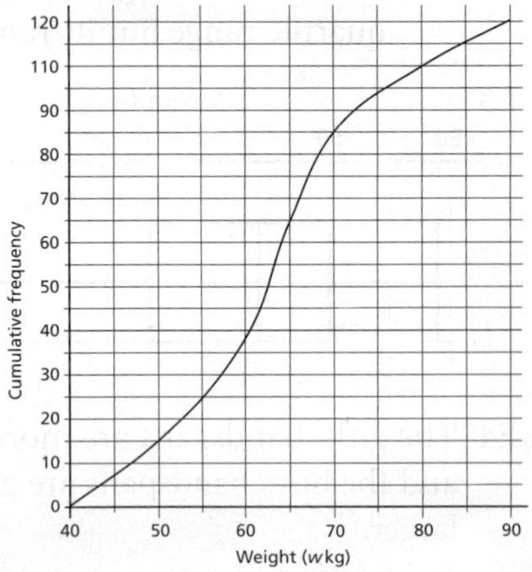

(c) 65kg

(d) Lower quartile = 57 kg
Upper quartile = 72 kg
Inter-quartile range = 15 kg

(e)

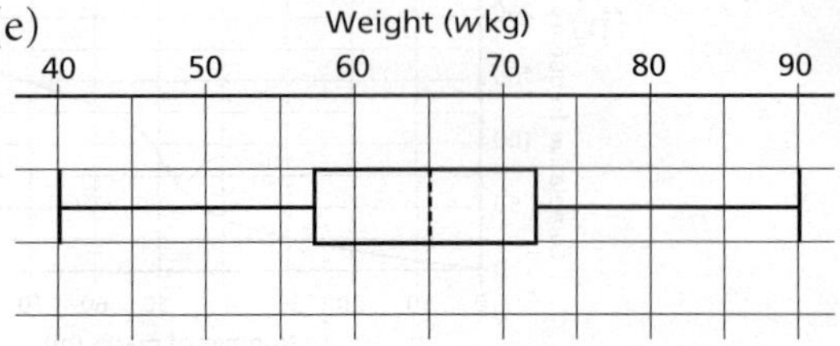

T2 (a) 25% (b) 25%

(c) Variety B is generally heavier than variety A (from comparing the medians). The weight of variety A apples is generally more variable than for variety B (from the interquartile range and the range).

T3 (a) About £60

(b)

Money received (£)	≤0	≤50	≤100	≤150	≤200	≤250	≤300	≤350
Cumulative frequency	0	32	44	48	50	51	51	52

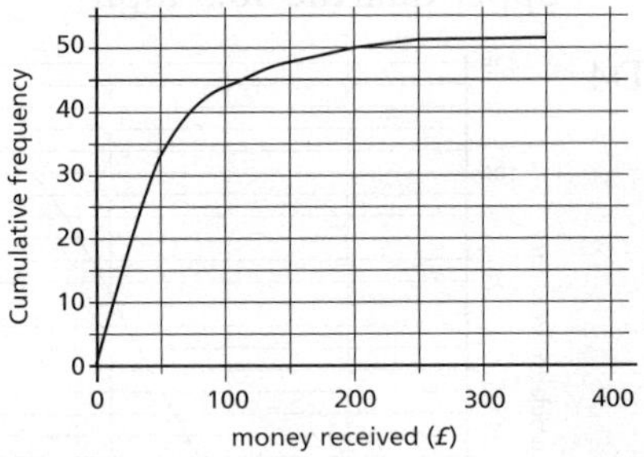

(c) Median = £40
Upper quartile = £70
Lower quartile = £20

(d) The student's explanation

Practice booklet

Section B (p 48)

1 (a) 38 (b) 42 (c) 43

2 (a)

Test result (r)	Cum. freq
$r \le 20$	3
$r \le 40$	10
$r \le 60$	30
$r \le 80$	62
$r \le 100$	72

(b)

Weekly wage (£w)	Cum. freq.
$w \le 110$	5
$w \le 120$	15
$w \le 130$	38
$w \le 140$	54
$w \le 150$	62

3 (a) 44 million

(b) (i) 3 million (ii) 5 million

(iii) 34 million

Section C (p 49)

1 (a) 12 (b) 28 (c) 52

(d) 23 (e) 12 (f) 41

(g) 35

2 (a)

Temp T°C	Cum. freq.
$T \le 0$	3
$T \le 10$	86
$T \le 20$	282
$T \le 30$	362
$T \le 40$	365

(b)

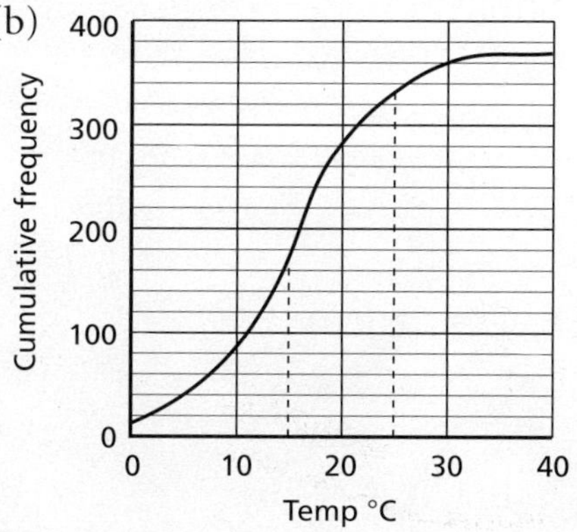

(c) 160 days (d) 12%

Section D (p 50)

1 (a) 33 (b) 21 (c) 42 (d) 21

(e) 21 marks (f) 110 students

2 (a)

Height	Cum. freq
$h \le 145$	3
$h \le 150$	8
$h \le 155$	18
$h \le 160$	30
$h \le 165$	45
$h \le 170$	51
$h \le 175$	54
$h \le 180$	56

(b)

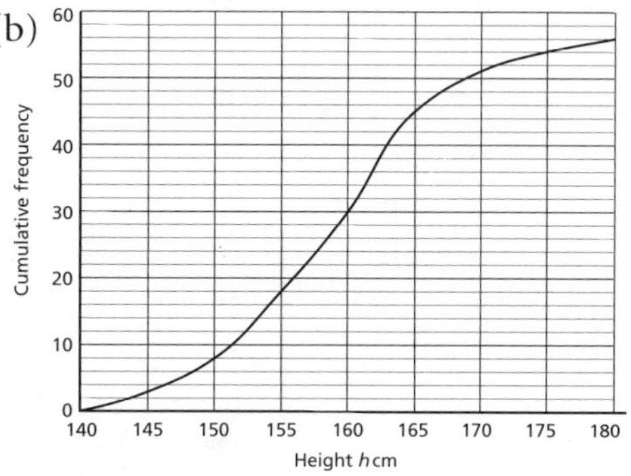

(c) (i) 159 cm

(ii) LQ = 154 cm UQ = 164 cm

(iii) 10 cm

3 Median speed ≈ 44 mph
Lower quartile ≈ 38 mph
Upper quartile ≈ 48.5 mph.

Section E (p 51)

1

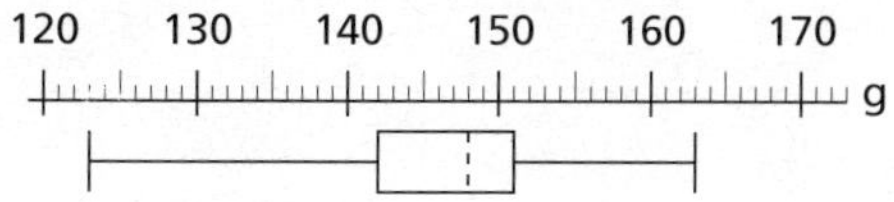

2 The student's comparisons

3 (a)

Test 1	Median	25
	LQ	20
	UQ	40
Test 2	Median	45
	LQ	30
	UQ	65

(b)

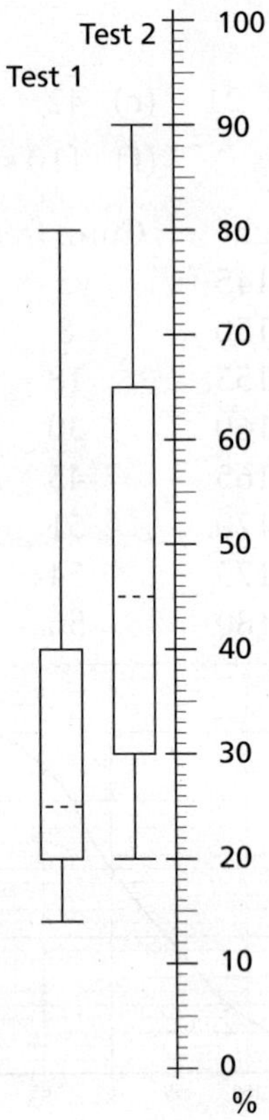

(c) Test 1 IQR = 20
Test 2 IQR = 35

(d) The student's comparisons

(e) Test 1: 10 passed
Test 2: 18 passed

15 *Simultaneous equations*

This is a long unit and you may well want to split it into two or more 'chunks'. One place to split the unit is after section D as sections A to D cover simultaneous equations where all coefficients are positive.

Practice booklet pages 52 to 55

A *What's the difference?* (p 131)

◊ Students should solve these informally by noticing the difference between the situations in each case.

◊ The problems can then be discussed as part of the introduction to section B. Equations could be formed and solved for each problem.

B **Solving equations** (p 132)

TG

◊ One approach is to think of the equations as balance puzzles and to 'spot the difference'. For example to solve $3a + 5b = 21$ and $3a + b = 9$, first note that you can obtain the left hand side of the first equation by adding $4b$ to the left hand side of the second equation. Then note that you can obtain the right hand side of the first equation by adding 12 to the right hand side of the second. Hence $4b = 12$ and the solution can be found.

By the end of the discussion should have moved away from the concrete to cover some examples with negative solutions.

◊ One way of generating equations is to ask two students to choose a number, write it down and keep it secret from other students. You look at the numbers and make up two true equations. Then the class tries to find which two numbers the students were thinking of.

◊ In your introduction, include simultaneous equations such as $a + b = 4$ and $3a + b = 2$. Students need to notice that you can obtain the left hand side of the second equation by adding $2a$ to the left hand side of the first equation and you can obtain the right hand side of the second equation by adding $^{-}2$ to the right hand side of the first equation giving $2a = {}^{-}2$ (not $2a = 2$). Students meet this 'problem' in question B5.

C **Shall I compare thee …** (p 133)

The first diagram leads to a problem where **one** of the equations has to be multiplied, and the second diagram leads to multiplying **both** equations.

TG

◊ Trying to find the prices of the different individual items leads to pairs of equations that cannot be directly compared. For example, the first diagram leads to $3c + b = 145$ and $5c + 3b = 275$ (working in pence). By the end of the discussion, students should understand that you can multiply both sides of an equation by the same number and still have a true statement about the unknowns. The first equation can be multiplied by 3 to obtain $9c + 3b = 435$ (thinking of the price of 9 coffees and 3 buns might help students see this) and now we can compare this with $5c + 3b = 275$ to give $4c = 160$ and $c = 40$.

For the second pair of pictures, **both** equations need to be multiplied - show two different ways of approaching this, for example multiplying by 2 and 3 or 5 and 2.

E *Spot the difference* (p 136)

This is the first section that deals with pairs of simultaneous equations where one of the equations involves subtraction.

It is a short section that is **not essential** but gives an alternative method to the adding approach that is considered in the next two sections. For example the problem in G3 (c) could be approached by multiplying the first equation by 2 and considering the difference. It can also help with pairs of equations that both involve subtraction (Section J). However, the section can be omitted if you don't think students will benefit from looking at a variety of approaches.

TG

◊ Ask students to form equations where p is Patsy's number and q is Quentin's ($3p + q = 40$ and $3p - q = 32$). Ask students what you need to add to the left hand side of the second equation to obtain the left hand side of the first. Hence obtain $2q = 8$ giving $q = 4$ and $p = 12$. Some students may see that you can 'add' the equations to obtain $6p = 72$ and they could be encouraged to compare both of these approaches.

H *Mixed problems* (p 139)

You could introduce this section by asking students to solve a set of mixed problems and discuss their methods.

I *To solve or not to solve?* (p 139)

TG

◊ Pose the problem at the bottom of page 139 and ask students to try and explain their findings as fully as they can.

The results are tabulated below:

Equations	Solution
A and B	$x = 1, y = 4$
A and C	$x = 1, y = 4$
A and D	$x = {}^-5, y = 7$
B and C	An infinite number of solutions of the form $x = a, y = 5 - a$
B and D	None
C and D	None

TG

Students are likely to find these results difficult to describe or explain. If they don't think of it for themselves, ask them to draw a graph for each equation on the same set of axes and to relate their results to the graphs. They should appreciate that, for example,

- Pairing B and C gives an infinite number of solutions illustrated by the fact that the equations give the same line
- Pairing C and D gives no solutions illustrated by the fact that the equations give parallel lines
- Pairing A and B **does** give a solution illustrated by the fact that the lines intersect and the coordinates of the point of intersection gives the solution.

This can be related to what happens when you try to solve the equations algebraically.

J *Further solving* (p 141)

This short section looks at pairs of equations that both involve subtraction. It also considers problems that can be solved by rearranging one of the equations. It does not explicitly mention solving by substitution.

TG

◊ Students can try the problems and discuss their methods of solution.

Equation A Looking at the difference (or formally subtracting), we can see that $x = 4$.

Equation B Either multiply first equation by 4 and then look at the difference to obtain $3p = 30$ or look at the difference to start with to obtain $3q = 6$. The second method needs more skill with negatives.

Equation C Either multiply second equation by 3 to obtain $13a = 39$ or multiply first equation by 5 and the second by 2 to obtain $13b = 13$. Again, the second method needs more skill with negatives.

Equation D Both equations need to be multiplied here, again either equating coefficients of h or k.

A *What's the difference?* (p 131)

A1 (a) 120 grams (b) 10 grams

A2 (a) 5 grams (b) 160 grams

A3 £0.95

A4 A bun costs 30p and a tea costs 80p.

***A5** ♥ = 10 ■ = 2 ◆ = 1

B *Solving equations* (p 132)

B1 (a) $a = 4, b = 1$ (b) $a = 1, b = 5$
(c) $a = 2, b = 2$ (d) $a = 3, b = 3$
(e) $a = 10, b = 7$ (f) $a = 5, b = 8$

B2 (a) C (b) $b = 55, e = 75$

B3 (a) $c = 0.5, d = 3$ (b) $h = 2, k = 1.5$
(c) $p = 1, q = 2.5$

B4 (a) $a = 2, b = {}^{-}1$ (b) $p = 4, q = {}^{-}1$
(c) $x = {}^{-}2, y = 10$ (d) $g = 4, h = 1$
(e) $m = {}^{-}3, n = 3$ (f) $c = {}^{-}0.5, d = 5$

B5 (a) $a = {}^{-}1, b = 7$ (b) $x = {}^{-}2, y = 4$
(c) $k = 4, h = {}^{-}3$

C *Shall I compare thee ... ?* (p 133)

C1 (a) $3a + 6b = 21$ (b) $a = 3, b = 2$

C2 (a) $6x + 4y = 32$ (b) $x = 2, y = 5$

C3 (a) $a = 4, b = 1$ (b) $p = 6, q = 2$
(c) $x = 2, y = 2$ (d) $g = 1, h = 1$
(e) $m = 3, n = 0.5$ (f) $c = 5, d = {}^{-}1$

C4 (a) $6m + 15n = 27$
(b) $6m + 4n = 16$ (c) $m = 2, n = 1$

C5 (a) $4x + 10y = 28$ (b) $15x + 10y = 50$
(c) $x = 2, y = 2$

C6 (a) $v = 4, w = 0$ (b) $p = 4, q = 1$
(c) $m = 1, n = 1$ (d) $k = 4, h = 3$
(e) $x = 2, y = 5$ (f) $a = 3, b = 1$

C7 (a) $v = {}^{-}2, w = 5$ (b) $p = 2, q = {}^{-}1$
(c) $m = {}^{-}2, n = 8$

D *Forming and solving equations* (p 135)

D1 (a) (i) $4b + 3w = 58$
(ii) $5b + 6w = 86$
(b) A black brick weighs 10 grams.
A white brick weighs 6 grams.

D2 $2p + q = 90$
$p + 3q = 95$
A candy bar costs 20p.

D3 (a) $8x + 2y = 19$ (b) $4x + 6y = 22$
(c) $x = 1.75, y = 2.5$

D4 Using c to stand for the cost of a CD in £s and t for the cost of a tape in £s, the equations are:
$3c + t = 8.92$
$2c + 5t = 12.23$

The cost of a CD is £2.49

D5 A golden egg weighs 1 kg.
A goose weighs 3 kg.

D6 $x + y = 40$
$2x + 3.5y = 92$

$x = 32$ and $y = 8$ so Greg sold 32 tickets at £2.00 each and 8 tickets at £3.50 each.

E *Spot the difference* (p 136)

E1 (a) $3y$ (b) $x = 1, y = 3$

E2 (a) $p = 5, q = 2$ (b) $a = 4, b = 3$
(c) $c = 3, d = 3$ (d) $x = 10, y = 2$
(e) $h = 2, k = 4$ (f) $m = 40, n = 39$

F *It all adds up* (p 137)

F1 (a) $p = 3, q = 4$ (b) $a = 5, b = 3$
(c) $c = 1.5, d = 1$ (d) $x = 5, y = 2$
(e) $h = 11, k = 2.5$ (f) $m = 3.5, n = 0.5$

F2 $a + 2b = 11$
$a - 2b = 3$

$a = 7, b = 2$

F3 (a) $p = 4, q = {}^{-}1$ (b) $a = 4, b = {}^{-}2$
(c) $c = 5, d = {}^{-}0.5$

G *Multiply to eliminate* (p 138)

G1 (a) $4x - 2y = 4$ (b) $x = 3, y = 4$

G2 (a) $9a + 6b = 39$ (b) $a = 3, b = 2$

G3 (a) $v = 5, w = 3$ (b) $p = 14, q = 5$
(c) $m = 3, n = 0.5$ (d) $k = 3, h = 0.5$
(e) $x = 5, y = {}^{-}1$ (f) $a = 3, b = 0.25$

G4 (a) $5x - 10y = 10$ (b) $4x + 10y = 44$
(c) $x = 6, y = 2$

G5 (a) $p = 1, q = 1$ (b) $a = 3, b = 2$
(c) $c = 3, d = {}^{-}1$ (d) $x = 4, y = {}^{-}1$
(e) $h = 1, k = 2$ (f) $m = 5.5, n = 1.5$

H *Mixed problems* (p 139)

H1 $x = 6, y = 2$

H2 $x = 3, y = {}^{-}2$

H3 $x = 4, y = {}^{-}1$

H4 (a) $x = 3.5, y = 0.5$
(b) $a = 2, b = {}^{-}1$
(c) $c = 3, d = 0.25$
(d) $x = 3.5, y = {}^{-}2$
(e) $h = {}^{-}1, k = {}^{-}2$
(f) $m = 2.75, n = 0.625$

I *To solve or not to solve?* (p 139)

I1 (a) (i) $x = 1, y = 3$ (ii) $x = 0, y = 4$
(b) The lines are parallel and lines must meet for the simultaneous equations to have a solution.

I2 (a) Graphs leading to $x = 4, y = 2$
(b) Graphs leading to $x = 2, y = 1$

I3 (a) (i) $x = 3, y = 7$ (ii) $x = 2, y = 5$
(iii) $x = 1, y = 3$ (iv) $x = 8, y = 2$
(b) The lines are parallel and lines must meet for the simultaneous equations to have a solution.

I4 (a) Graphs leading to $x = 1.5, y = 3.5$
(b) Graphs leading to $x = 11, y = 4$

J *Further solving* (p 141)

J1 (a) $p = 3, q = 1$ (b) $a = 1, b = 0.5$
(c) $c = 6, d = 5$ (d) $x = 8, y = 3$
(e) $h = 2, k = {}^{-}1$ (f) $m = 4, n = 2$

J2 (a) C (b) $x = 3, y = 7$

J3 (a) $p = 6, q = 1$ (b) $a = 5, b = 2$
(c) $c = 7, d = 3$

Test yourself (p 142)

T1 (a) 70p (b) 50p

T2 $x = 3, y = {}^{-}0.5$

T3 $x = {}^{-}0.5, y = 3$

T4 $2s + l = 758$
$4s + 3l = 1882$
A small mug weighs 196 grams.
A large mug weighs 366 grams.

T5 50p

T6 $x = 3.5, y = 5.5$

T7 $x = 3, y = {}^{-}1$

T8 $x = 3, y = {}^{-}2$

T9 $x = 0.75, y = {}^{-}0.5$

Practice booklet

Sections A, B and C (p 52)

1 35p

2 ■ = 5, ▲ = 2 ● = 3

3 (a) $x = 2$ $y = 3$
(b) $x = 3$ $y = 1$
(c) $x = 4$ $y = 3$

4 (a) B (b) $c = £1.50$ $h = £2.00$

5 (a) $a = 4$ $b = 0.5$
(b) $x = {}^{-}2$ $y = 4$
(c) $h = 3$ $k = 1$
(d) $x = 8$ $y = {}^{-}2$
(e) $a = 1.5$ $b = 0.5$
(f) $m = 4$ $n = {}^{-}3$

Section D (p 52)

1 (a) (i) $4m + 3p = 745$
(ii) $7m + 6p = 1375$
(b) $m = 115$ $p = 95$

2 $3g + 5c = 335$
$9g + 3c = 705$
$g = 70$ $c = 25$

3 $3d + 2t = 27$
$5d + 4t = 49$
$d = 5$ $t = 6$

4 Jelly beans 130 g
Chocolate raisins 140 g

Sections E, F and G (p 53)

1 (a) $x = 3$ $y = 2$
(b) $a = 4$ $b = 3$
(c) $p = 25$ $q = 20$

2 (a) $x = 4$ $y = 1$
(b) $a = 5$ $b = 2$
(c) $p = 5$ $q = 2.5$

3 $2m + 3n = 21$
$3m - 3n = 9$
$m = 6$ $n = 3$

4 (a) $x = 4$ $y = 2$
(b) $a = 9$ $b = 6$
(c) $a = 2$ $b = 1$
(d) $x = 5$ $y = {}^{-}2$
(e) $f = 3$ $e = {}^{-}1$
(f) $x = 1.5$ $y = 0.5$

Section H (p 54)

1 (a) $x = 5$ $y = 3$
(b) $m = 2$ $n = {}^{-}3$
(c) $x = 3.5$ $y = 1.5$
(d) $a = 4$ $b = 2$
(e) $h = 0.5$ $k = {}^{-}1$
(f) $a = 0.4$ $b = 0.52$
(g) $p = 3$ $q = 8$
(h) $c = 10$ $d = 2$

Section I (p 54)

1 (a) $x = 3$ $y = 7$
(b) $x = 1$ $y = 3$
(c) $x = 11$ $y = {}^{-}1$

2 Graphs leading to:
(a) $x = 7$ $y = 5$
(b) $x = 0$ $y = 2$

3 (a) (i) $x = 3$ $y = 5$
(ii) $x = 0$ $y = {}^{-}1$
(iii) $x = 6$ $y = 2$
(iv) $x = 4$ $y = 1$
(b) The lines are parallel and lines must meet for the simultaneous equations to have a solution.

4 Graph leading to:
$x = 4$ $y = 7$

Section J (p 55)

1 (a) $a = 4$ $b = 8$
(b) $x = 7$ $y = 3$
(c) $m = 2$ $n = {}^{-}1$

2 (a) B
(b) $y = 11$ $x = 1$

3 (a) $x = 8$ $y = 1$
(b) $m = 10$ $n = 1$
(c) $h = 3$ $k = 4$
(d) $x = 5$ $y = {}^{-}2$

Review 3 (Student's book p 143)

1 (a) 4 : 1 (b) 20%

2 $3a + c = 12$
$2a + 4c = 13$
gives $a = 3.5$, $b = 1.5$ so the tickets cost £3.50 and £1.50 respectively

3 (a) About 75 students
(b) 50%
(c) 54
(d) Lower quartile is about 44
Upper quartile is about 63
(e) About $82\frac{1}{2}$%

4 1 : 50 000

5 (a) 3.0 cm (b) 70° (c) 8.2 cm

6 (a) £160 (b) 48%

7 (a)

Number of hours (h)	Cumulative frequency
$h \le 5$	2
$h \le 10$	7
$h \le 15$	21
$h \le 20$	39
$h \le 25$	71
$h \le 30$	86
$h \le 35$	89
$h \le 40$	90

(b) The student's cumulative frequency curve.
(c) About 21 hours
(d) In the survey 69 out of 90 people watched more than 15 hours of TV each week. This is 77% so the survey data does support this headline.

8 (a) 300 ml (b) 30 ml

9 $x = 5, y = {}^{-}2$

Mixed questions 3 (p 56)

1 (a) 20, 25 (b) £34, £25.50, £25.50

2 (a) 1 : 1.25 (b) 0.8 : 1

3 (a) 11.6 cm (b) 7.7 cm
(c) 55.3° (d) 51.5°

4 1.8 m

5 (a)

≤ 1	≤ 1.5	≤ 2	≤ 2.5	≤ 3	≤ 3.5
7	23	62	90	108	110

(b)

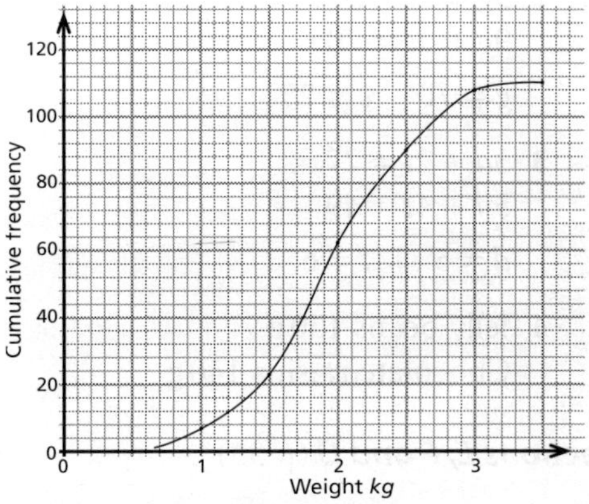

(c) 7.3%
(d) (i) 1.9 (ii) 1.6, 2.3–2.4
(iii) 1.6 to 2.3
(d)

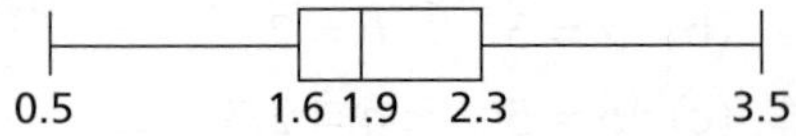

6 (a) $x = 5, y = 3$ (b) $x = 4, y = {}^{-}2$
(c) $x = {}^{-}2, y = 4$

7 $x = 4, y = 4$